Water and Urbanism in Roman Britain

The establishment of large-scale water infrastructure is a defining aspect of the process of urbanisation. In places like Britain, the Roman period represents the first introduction of features that can be recognised and paralleled to our modern water networks. Writers have regularly cast these innovations as markers of a uniform Roman identity spreading throughout the Empire and bringing with it a familiar, modern, sense of what constitutes civilised urban living. However, this is a view that has often neglected to explain how such developments were connected to the important symbolic and ritual traditions of waterscapes in Iron Age Britain.

Water and Urbanism in Roman Britain argues that the creation of Roman water infrastructure forged a meaningful entanglement between the process of urbanisation and significant local landscape contexts. As a result, it suggests that archetypal Roman urban water features were often more closely related to an active expression of local hybrid identities rather than aligned to an incoming continental ideal. By questioning the familiarity of these aspects of the ancient urban form, we can move away from the unhelpful idea that the Roman precedent is a central tenet of the current unsustainable relationship between water and our modern cities.

This monograph will be of interest to academics and students studying aspects of Roman water management, urbanisation in Roman Britain and theoretical approaches to landscape. It will also appeal to those working more generally on past human interactions with the natural world.

Jay Ingate is currently a sessional lecturer at Canterbury Christ Church University, UK. He was awarded his PhD by the University of Kent, UK in 2014. He has written articles on the interpretation of aqueducts in Roman Britain, the development of Roman London's waterscape and post-human approaches to the Roman world.

Studies in Roman Space and Urbanism
Series editor: Ray Laurence
Macquarie University, Australia

Over the course of the last two decades the study of urban space in the Roman world has progressed rapidly, with new analytical techniques, many drawn from other disciplines such as architecture and urban studies, being applied in the archaeological and literary study of Roman cities. These dynamically interdisciplinary approaches are at the centre of this series. The series includes both micro-level analyses of interior spaces and macro-level studies of Roman cities (and potentially also wider spatial landscapes outside the city walls). The series encourages collaboration and debate between specialists from a wide range of study beyond the core disciplines of ancient history, archaeology and Classics, such as art history and architecture, geography and landscape studies, and urban studies. Ultimately the series provides a forum for scholars to explore new ideas about space in the Roman city.

Water and Urbanism in Roman Britain
Hybridity and Identity
Jay Ingate

For further information about this series please visit www.routledge.com/classicalstudies/series/SRSU

Water and Urbanism in Roman Britain

Hybridity and Identity

Jay Ingate

Routledge
Taylor & Francis Group

LONDON AND NEW YORK

First published 2019 by Routledge

2 Park Square, Milton Park, Abingdon, Oxon OX14 4RN
605 Third Avenue, New York, NY 10017

Routledge is an imprint of the Taylor & Francis Group, an informa business

First issued in paperback 2021

Publisher's Note

The publisher has gone to great lengths to ensure the quality of this reprint
but points out that some imperfections in the original copies may be apparent.

British Library Cataloguing-in-Publication Data
A catalogue record for this book is available from the British Library

Library of Congress Cataloging-in-Publication Data
Names: Ingate, Jay, author.
Title: Water and Urbanism in Roman Britain: Hybridity and
Identity/Jay Ingate.
Description: Landon: New York: Routledge, Taylor & Francis
Group, 2019. |
Series: Studies in Roman space and urbanism | Includes
bibliographical references and index.
Identifiers: LCCN 2018057715 (print) | LCCN 2018057953
(ebook) | ISBN 9781315206707 (ebook) | ISBN 9781351797849
(web pdf) | ISBN 9781351797825 (mobi/kindle) |
ISBN 9781351797832 (epub) | ISBN 9781138634695 (hardback:
alk. paper)
Subjects: LCSH: Municipal water supply–England–History–To
1500. | Urbanization–England–History–To 1500. | Great Britain–
History–To 449.
Classification: LCC TD257 (ebook) | LCC TD257.I54 2019 (print)
| DDC 363.6/109362–dc23
LC record available at https://lccn.loc.gov/2018057715

ISBN: 978-1-138-63469-5 (hbk)
ISBN: 978-1-03-217827-1 (pbk)
DOI: 10.4324/9781315206707

Typeset in Sabon
by Wearset Ltd, Boldon, Tyne and Wear

For my Mum

Contents

Figures

Acknowledgements

This book is largely a product of research undertaken for the completion of my PhD at the University of Kent between 2009 and 2014. As such, it could not have been completed without the departmental scholarship awarded to me by the School of European Culture and Languages. Particular thanks must be given to my PhD supervisor Steve Willis, who gave insightful feedback on many of the ideas that form the core of this book. My examiners J. D. Hill and Ray Laurence also made key suggestions that would help clarify some of the main points of the research. As the series editor of *Studies in Roman Space and Urbanism*, Ray Laurence has also aided in the conversion of my PhD thesis into this book, giving me timely support and feedback from submission of the proposal through to the production of a final manuscript.

As part of the process of assembling the research for this book, I contacted a number of individuals with specific enquiries. Some of these contacts were made several years ago now, but my thanks for the help: Ellen Swift, Patty Baker, Richard Brickstock, John Casey, Matt Edgeworth, Tom Moore, Simon Pratt, Richard Reece, Adam Rogers, Jake Weekes, Simon West and Kenneth Sheedy. There are also many people with whom I have informally discussed this research over the last few years – thank you for enduring my monologues on water!

I should also note here that since 2015 I have been employed as a sessional lecturer at Canterbury Christ Church University. While this is a post without research duties, my colleagues have been as considerate as possible in distributing teaching responsibilities so as not to disadvantage my research activities. Between 2014 and 2018 I also worked part-time for the Northampton Healthcare Foundation Trust and greatly appreciated their organisational flexibility, which allowed me to continue my academic pursuits while earning a consistent wage.

Finally, my family have been unquestioningly supportive throughout the completion of my PhD, and in the years that have followed while producing this work. Without their encouragement and assistance, I would not have been able to finish this book.

Chapter 1

Water and urbanism

Introduction

Water infrastructure is a familiar part of any consideration of Roman towns and cities. Indeed, our views on the subject are a significant contributory factor to the perception of a close alignment between Roman urbanism and our modern urban identity. Today, the presence of 'running water' is often put forward as a defining characteristic of prosperous nations, and something that separates them from the struggles of the so-called 'developing world'. Likewise, it has been common practice to treat the Roman establishment of urban water supply as a seed of modern civilisation that culturally elevated the people of ancient Europe to something resembling our lifestyle. The success of Roman aqueducts, wells, sewers and bathhouses has therefore chiefly been measured by the extent to which they echo the technological sophistication at the heart of modernity. It is thus a topic that treasures similarity and familiarity, something which is of particular importance within provincial locations such as Roman Britain. Here, archaeologists have been satisfied to explain the construction of such features as logical parts of urban improvement, unrelated to previous traditions and requiring little in the way of justification for their creation in the Roman period. They remain exemplars of the popular, but rather colonial, refrain about 'what the Romans did for us'.

In recent times Roman archaeologists have sought to distance themselves from such interpretations of the past. Smaller examples of material culture have been championed as more valuable in discovering the 'everyday lives' of people in the past. One might say that the topic of urban water management has thus become somewhat unfashionable. At the very least, it is seen as a 'known' subject with comparatively little room for innovative theoretical approaches or interpretations. This is why the engineering of these structures has become such focus for scholars – it is seen as a direction that can provide fresh insight. Yet it is also an approach emphasising a role for water in urban centres that speaks to twentieth-century rationality, rather than exploring the breadth of the ancient associations. There is no intention in this book to provide another copious technical treatise on the practicalities of water supply in the Roman world. Instead, it is a discussion of that broader meaning of water for communities in the past, and what this can tell us about changing urban identities and perceptions. Of course, structures such as wells and bathhouses, in particular, are represented in a wide array of settlements. But it is the focus of this book to pinpoint water's impact on the experience of the so-called 'most Roman' places in the province, tracing their transition from the prehistoric settlements into Roman towns, with a large emphasis falling on the developments of the first and second centuries AD. Accordingly, the settlements under scrutiny (Figure 1.1) have been identified as the principal urban centres of Britain by authors

Figure 1.1 Locations of British towns analysed in this book.

Source: Drawn by author.

like John Wacher (1975, 1995). As Rogers (2008, 2011a) pointed out, these 21 settlements also provide the best archaeological evidence from which one can interpret urban change.

Too often, past approaches to this subject have devalued the actual importance of the water itself, with writers assuming that it can be as easily quantified as the familiar homogeneous resource of our modern cities. This book gives evidence suggesting the contrary in the ancient Roman world: it was a time when water was powerful and meaning-laden, with a complex set of symbolic and ritual associations across the many different cities of the Empire. It is proposed that even in Italy, at the heart of the Roman world, this localised significance was a key consideration when establishing a water supply. While there were practical considerations, management of water was undeniably a way to shape the experience of urban space actively by utilising this historically charged and symbolically profound medium. Of course, by acknowledging complexity in the implementation of water networks in Italy, we are compelled to re-examine the evidence of provincial locations such as Roman Britain. Within these temperate European settings there was also a rich pre-Roman tradition of water manipulation for symbolic and ritual reasons. These practices have been largely ignored when considering Roman-period developments, even when they occur within the same locality. This book aims to show how such circumstances could have provided diverse motives for communities to establish urban water networks. By engaging with such ideas we can gain an appreciation of the extent to which our familiar conception of Roman urban space belies the strange hybrid identities of cities and towns in provinces like Britain.

To emphasise the potential in such new approaches it is necessary to analyse the facets of water that have been most closely aligned with dialogues of familiarity and generalisation. Adam Rogers (2013) convincingly argued for water to be recognised as meaning-laden within Romano-British towns, principally by discussing the long-term changes and interactions with 'natural' waterscapes. However, hitherto analysis of this ilk has stopped short of acknowledging the extent to which these approaches alter our view of the Roman urban water supply, which was often sourced through wells and aqueducts, and its main beneficiary, the bathhouse. These features are presented primarily as humanised pillars of practical, technologically advanced, Roman endeavour. As a result, even if we interpret further meaning (more common in the case of wells), this rarely contributes to wider progressive debates that highlight links between prehistoric British context and the later Roman urbanised environment (Hingley, 2005; Creighton, 2006; Mattingly, 2007; Rogers, 2011a). Understanding the hybrid complexity of these waterworks and defamiliarising or 'making strange' their function can greatly alter our conception of urban space as a whole.

The utility of defamiliarisation or 'making strange' was first explored by Shklovsky ([1917] 1965) in analysing the differences between poetic and everyday language. He argued that embracing complexity and new perspectives allows us to escape the 'habitualisation' of daily life. The act of 'making strange' frees us from the shackles of habitual perception and interpretation. This concept has been engaged with by some prehistorians, such as Richard Bradley (2007), but is largely at odds with the way we have approached Roman-period evidence. Nevertheless, a primary theme of this book is to explore how an appreciation of the complexities of water, in particular, can defamiliarise some of the archetypes of the Roman town.

Of course, it might be suggested that the traditional, familiar conceptions of Roman urban space are the best source of relevance for the modern world; indeed, this is the direction taken by many popular books and television programmes on the period (see, for instance, Mount, 2010). However, a greater relevance for twenty-first-century urban centres may be found in this process of defamiliarisation, or 'making strange'. Climate change means our future relationship with the natural world will be markedly different from that which characterised the twentieth and early twenty-first centuries. The treatment and usage of water within urban space will be a prominent part of this shifting of attitudes. Change is essential, but the status quo is often portrayed as the teleological directive of ancient precedent (see Fagan, 2012). If we proclaim the 2,000-year genesis of a particular approach to water, it becomes hard to convince people that a different approach is needed, or indeed possible. However, by acknowledging the real complexity of difference in our urban heritage, it is possible to see how the future relationship between our urban centres and the natural world could be based on a different logic. In fact, the basis of past connections to water outlined in this book may hold more in common with the proposals of new environmental writers than with the cultural baggage of the previous 100 years.

This chapter serves as an introduction and a critique of how Roman archaeologists have interpreted urban water infrastructure in the past, underlining how these views have often been deeply related to the pervading attitudes of mastery and marginalisation of water in modern cities. The intention is to show that while these views are largely accepted, they ignore a proportion of the classical literature and the archaeological evidence. It is also striking how, on this particular subject, the views of Roman archaeologists in Britain differ significantly from the ongoing work of prehistorians analysing the evidence of the Iron Age period that immediately preceded the Roman conquest. This chapter outlines the central themes of the book, namely that water was a medium which had a crucial role in shaping and remaking local identities, and therefore had a profound impact on how the Roman town was experienced. Moreover, by analysing water we can see the patchwork of hybrid identities within urban centres that both embrace incoming associations and continue to express long-lived connections to the landscape. While in our modern towns we are increasingly unaware of water, the people of Roman Britain had intimate knowledge of its presence in the landscape, and this helped to define their sense of urban space.

Water and twentieth-century approaches to Roman urbanism

The study of Roman urbanism has often been drawn towards monumental archetypes: the baths, fora and theatres have served to define towns as 'Roman' as well as dictating their relative economic and political import-ance within the Empire. This is illustrated well in the work of both Wacher (1975, 1995) and Rivet (1958) – authors who have become gateway sources for anyone wishing to gain a general understanding of urban life in Roman Britain. The analysis of Roman water networks has satisfied this traditional pursuit of elements of imperial largess and monumentality. The scale of these systems and the ingenuity required to make them function correctly are vivid examples of technological sophistication, a concept inseparable from our modern perception of civilisation. Thus it is easy to interpret the rationale behind their creation as an unproblematic perk of imperial occupation.

In this regard, the provision of a water supply has been portrayed as a functional requirement for the full transformation to a Roman way of life (see Mattingly, 2007: 280). Urban water structures have been seen as the primary 'enabling' feature that moved civilisation 'forward'. Their appear-ance in Roman towns often seems to be cast against the lack of such struc-tures in earlier history, thus underlining a supposed epoch change. We have even tried to explain the fall of the Roman Empire as a miscalculation in such systems, making them out of lead (Waldron, 1973; Nriagu, 1983). Here they are shown almost as the alpha and omega; the beginning and end to a civilised way of life that was only rediscovered in modernity. There is even a sense that we can question the 'Roman-ness' of British towns based on the technological sophistication of their water supply.

In many ways the subject has been bound to colonial discourse. As a vast and enduring Empire, Rome is preeminent in the annals of Western history. Its geographical coverage meant that many European powers of the modern age were, at one point, within its sphere of influence. The direct impact of the Empire upon these territories has resulted in a constant reference to Rome within their dialogues of national growth. At times this has been negative, perhaps best seen in France (King, 2001) with its post-1789 revolutionary spirit; but also overwhelmingly positive in the embrac-ing of notions of Rome in watershed moments of European imperialism in the nineteenth and twentieth centuries. Naturally, Italy has always had an active link with its past, but ideas of identification with 'Rome' were particularly strong during the Fascist period. Mussolini cultivated a sense of continuation from the ancient past that expressed itself vividly in town planning (Laurence, 1994a; Terrenato, 2001). However, to this day the unlikely rise of Britain to world superpower status debatably provides the closest parallel to Rome's similar ascension from its humble roots as a small city-state. It is unsurprising, then, with such comparison, that Britain (at its imperial zenith) also adopted a very positive view towards the Roman Empire and its achievements.

Britain's imperial elite was exposed to classical culture from a young age, with Latin literature being a principal focus of the public school education system (Vance, 1997; Stray, 1998; Hingley, 2000). Furthermore, the 'Grand Tour' had been a tradition for the British aristocracy throughout the seventeenth and eighteenth centuries. While initially this may have produced just an artistic affinity, rather than political agreement, it would soon lay the foundation for a more comprehensive identification with the Roman Empire. Once British influence had spread across the world and laid claim to a collection of provincial territories (making Victoria an empress in the process), there was a definite change in attitude (Hingley, 2001: 27). Part of this inevitably came from a justification of the aggressive colonial policy that had been adopted. The morally dubious actions that are part of imperial expansion were portrayed as 'progress' by using the Roman example. There was an effort to cultivate a structured sense of order on similar terms to that of Rome in the High Imperial period; the Pax Britannica was the successor to the Pax Romana (Hingley, 2005: 21). Essentially, Britain envisioned itself bequeathing civilisation to its conquered territories, reproducing the 'gift' Rome had made to Western Europe in antiquity (ibid.: 27). Technology and leisure were pillars in connecting the past to the present; this is perhaps best illustrated in the conscious aping of Roman aqueducts in the construction of the numerous railway viaducts of the Victorian age. The railways were, likewise, a great 'gift' of imperial occupation (Figure 1.2).

This teleological conception of history made it possible to place 'civilising' aims at the heart of the aggressive imperialism undertaken in places like India. The feelings of cultural proximity to the Roman period were even seen in the popular literature of the time. Hingley (2000) highlighted how famous writers like Rudyard Kipling often reinforced a positive connection between the Roman era and the British Empire. For instance, in the collection of short stories entitled *Puck of Pooks Hill* (Kipling, 1906) we have direct crossovers between the Roman period and Victorian England. Within this collection, the story entitled 'A Centurion of the Thirteenth' starts with a child pretending to be a Roman soldier in the woods, only to fire her handheld catapult and be confronted with an actual

Figure 1.2 The Welland viaduct in Northamptonshire, showing the similarity of its form to classical aqueducts.

Source: Picture taken by author.

centurion. The tone is light-hearted, but the centurion comes across as civilised and similar in his worldview to a modern Victorian man. Much is made of the similarity in geography, concerning the location of particular places in the past and present. Furthermore, the emphasis is upon this soldier, despite being a 'Roman' citizen, actually being a native of Britain; thus showing first hand the passing on of 'civilisation'. Some may suggest that Kipling is outlining a degree of entanglement that questions definitive identities. However, as with his poem 'The Roman Centurion's Song' (Kipling, 1911), the mixing of identity is between Roman Britain and Kipling's contemporary Britain. The historical denizens of the island are in fact depicted as 'native troops to drill', so just beginning their quest to embrace Roman identity and eventually forge the empire of Kipling's age.

This general cultural milieu is vital to note, because it was at this time that archaeology started to emerge as an academic discipline. In Britain there are probably few archaeologists who have been as influential as Francis Haverfield, the late Victorian/Edwardian scholar and Camden Professor at Oxford. Haverfield's work drew heavily upon the issues of colonial thought mentioned above. While Theodor Mommsen (1885) may have invented the idea (albeit in another context), Haverfield (1912) crystallised the concept of Romanisation in his publication *The Romanization of Britain*. It is no exaggeration to label this as one of the most influential contributions in the study of Roman Britain, and maybe the Western Roman provinces in general. 'Romanisation' described a conversion process that represented an ideal outcome for colonial occupation. It had tremendous relevance at the time it was written, with many historical scholars comparing the situation in India with the Roman provinces (Cromer, 1910; Lucas, 1912; Bryce, 1914).

As a concept, Romanisation endorses a teleological progression from native barbarian to civilised Roman. In this sense it is a theory which has functioned well in our compartmentalisation of history; it explains the transition from the Iron Age to the Roman period. Furthermore, it is a process that can be explained from an archaeological perspective, through the rapid adoption of Roman-style material culture and the appearance of distinct urban features like the aqueducts, bathhouses and sewers. Such new building projects are frequently depicted as the hallmarks of this inevitable cultural revolution. Mattingly (2007: 376) summarises this sentiment with the dry observation that surely everyone would want to live in a villa instead of a roundhouse. Haverfield (1913: 14) underlined the simplicity of this choice by proclaiming that the use of the square and straight lines was a critical attribute which separated civilised man from barbarity. For most of the last century our understanding of urban centres has been cradled in this type of caricature, with central amenities forming a particular focus. Buildings like bathhouses are seen as instant markers of progress, indicating a 'Romanising' population who were striving towards a (modern?) civilised ideal. This is acutely expressed in the definition archaeologists have given to the Roman town, identifying it as a place that provided administration, trade, amusement and protection (Wacher, 1995: 257) – a list that falls comfortably within the definition of a modern town.

This connection between urban centres of the Roman period and their modern equivalents has been replicated in economic terms. Rostovtzeff (1926), for example, directly associated the ancient town with the modern commercial centre. Even the work of Weber (1958, 1978), Finley (1973) and Jones (1974), which rejected this strict adherence to direct implementation of modern economic models in ancient contexts, was based on the primacy of a 'consumer city' (Kaiser, 2000). According to these accounts, the purpose of the town was to organise and exploit the countryside for economic gain. It is a theoretical approach that has been influential regarding water management, casting it as exploitative of the broader context of urban centres. So while there may have been a move away from strict modern economic principles (particularly by Finley), there was still a move to define Roman towns in primarily recognisable economic terms. While this was to some extent true, it nevertheless gave a somewhat one-dimensional meaning to these settlements (Rogers, 2008: 53). Also, such weighty contributions were often based upon the copious evidence of the Mediterranean heartland, rather than of the provincial extremes such as Britain, which led to more acute generalisation in work on the northern provinces.

Rather than explaining the reasons for variability or similarity, there was an effort to force these models on to provincial archaeology. Instead of exploring differences, simple functions were seen to be indicated by (but not limited to) particular structures: the forum provided administration; the theatres and bathhouses were sources of entertainment; the markets and shops were evidence of trade; plus the city walls offered protection. As Kaiser (2000: 6) notes, this change in thought meant that a settlement could still be deemed a 'Roman' town despite lacking the traditional markers; so, for example, the absence of a curia did not necessarily mean there was no city council. This meant we could generalise sites despite certain incongruities in the archaeological record. As is shown later in this book, the bathhouse is a key example of this, with fragmentary evidence often sufficient to generate grand interpretations along Mediterranean lines. In effect this was more of a proving exercise, to illustrate sufficient 'Romanisation' to be relevant in the overall theoretical dialogue.

However, the real effect of this was that archaeologists started to produce studies that could imprint large-system thinking upon Roman towns in Britain, creating a scientific logic to the establishment of all settlements. There is no doubt that post-war economic theory reinforced a positive approach to Roman studies in Italy. Indeed, Terrenato (2001: 80) notes how the framework laid down by Mussolini (proclaiming the primacy of Rome in Western culture) was still utilised, but with aspects of ancient Italian economic power replacing the spiritual inheritance. In Britain, without the same wealth of archaeological resources, such theories were taken to their systematic extreme. Hodder and Hassall (1971) are a prime example of this approach, creating a pseudoscientific system that visually represented Roman towns and their territories. It was a form more akin to an illustration of a cellular structure in biology, with cities seen as performing a role in a local milieu while being a standardised element in a wider whole. This processualist perspective echoed the impact of functionalism in other academic disciplines at this time. The aim was to show the economic distribution and precision that characterised the siting of these towns within civitas resource zones. In this regard, the economic rationale further consummated the familiarity of the Roman town by firmly anchoring its creation and purpose in modern theoretical systems. No doubt the Roman Empire and its various settlements possessed strong economic and administrative links, but to make such a bold declaration of primary purpose underlined a prevailing one-dimensional view in our perception of the past. This perspective is especially relevant considering the genesis of economic theory itself, forged as a companion (or reaction) to the capitalist systems of modernity. There is naturally a disagreement about whether one can apply such a theory to a non-capitalist society of the past (Morley, 2004: 34). Indeed, while these studies may have seemed like a radical departure from previous dialogues of urban growth, they only served to maintain the status quo of modern projection when interpreting the evidence of the Roman period.

In many ways the momentum of archaeology's growth and evolution has helped to solidify a certain perception of the Roman Empire. This is particularly marked in Britain thanks to the dual aspects of our immediate imperial legacy plus the quest for relevance and respect for the discipline as a whole. Despite the dissolution of the European empires as the last century progressed, the foundations of this cultural proximity were already deeply entrenched within our society and academia itself. As such the ideals of processual archaeology merely furthered a familiarity between the Roman town and its modern equivalent. Our Western civilisation has been defined by its urban amenities, and therefore, following the above logic, the Roman town must have had similar attributes. With this modern point of view, anything potentially unusual or divisive in the pattern has been labelled as 'native' or rationalised to fit the overall picture.

This has become increasingly problematic over the last 20 years as the wider theoretical dialogue in Roman archaeology in Britain has sought to distance itself from such approaches (Hingley, 2000, 2001, 2005; Creighton, 2006; Mattingly, 2007; Rogers and Hingley, 2010). Consequently, a situation has arisen where studies of urban water have become isolated from the broader debates on the subject, with no really satisfactory explanation for the uptake of these networks in provinces like Britain beyond outdated colonial approaches. The study area has thus turned inward and looked to technical elements of these structures. Indeed, Hodge (1992: 4) lamented the lack of genuine technical analysis on the workings of aqueducts, chiding the past efforts of archaeologists. However, 27 years on from that contribution it seems there could be a similar sentiment expressed about the technical perspectives which now dominate studies of water. While the emphasis on technical detail may appear to be theoretically neutral, the reality is that by concentrating on such elements we only serve to bring these structures further into the rationale of the twenty-first century, detaching them from their actual context.

Justifying water networks

The theoretical abstraction of water from its contextual meaning has led discussions of Roman urban water features to focus on issues of scale.

If people see water as merely a homogeneous resource for a 'consumer city', then the only real areas of interest lie in how much of this resource is being extracted and used. This way of thinking ends up being far more concerned with the structures bearing water than with the water itself. In this regard, perhaps the most startling aspect regarding the analysis of features like Roman aqueducts is the lack of attention we have given to the actual water. This may seem a rather paradoxical comment considering the nature of the subject. However, while many archaeologists and engineers have given us wonderful and copious accounts of the structural elements of aqueducts, water has been a statistic needed to fulfil an equation. The high watermark of such an approach might be Kleijn's (2001) work on the water supply of Rome: the author attempts to use the relative capacities of the aqueducts that led into Rome to make estimates about the population of the city. It is an engaging account which has contributed to understanding of the technological sophistication of such engineering in the capital, but it also overemphasises a functionalist purpose for urban water supply. It assumes that in the Roman world water management was understood in twentieth/twenty-first-century terms, with the proportion of water supplied directly correspondent to individual requirements.

Of course, the dangers of emphasising pragmatic interpretations of water supply have been recognised by other writers. For example, Hodge (1992: 5) was quick to chide the 'social historian' for putting an essentialist spin on the aqueducts of Rome; he sees them as an expression of luxury and conspicuous consumption. Certainly, when we look at the total water supply provided by Rome's first nine aqueducts, a figure that could have been around $600,000\,m^3$ per day (Kleijn, 2001: 58), we can see the merit of such an interpretation. Considering the proliferation of additional wells and water sources that would have been a feature of any ancient city (Hodge, 1992: 5), this figure is colossal. The cost of constructing such structures would have been vast, estimated at up to 3 million sesterces per kilometre (Leveau, 1991; Chanson, 2004); to give this some context, in the first century AD the average Roman solider earned approximately 1,600 sesterces a year (Southern, 2007: 108). Combine this cost with the need for regular upkeep, and there would have been a massive amount of money outlaid. It is undeniable that a water network of this scale was a product of wealth, power and largess. The immediate association we make with such activity is the decadent elite lifestyle often portrayed as characteristic of the Imperial period. But if aqueducts were just a product of this flagrant display of wealth, it seems odd that we hear no voice of opposition to them. There was a tendency on the part of some writers of the Imperial period to portray this decadent lifestyle as a negative development away from a golden Republican era. By contrast, in the case of aqueducts the Imperial-period Aqua Virgo is often mentioned alongside the High Republican Aqua Marcia (Statius, *Silvae*: 1.5), with both treated equally.

In addition, the aqueducts seem to be free from any negative association with so-called 'bad' emperors. While Nero's Domus Aurea was seen as an embarrassment after *damnatio memorae*, his extensive reworking of the Aqua Claudia (which provided water to the Domus Aurea) did not have any negative repercussions.

With this in mind, it is interesting to analyse our thought process. The concept of 'luxury' is an outcome that naturally aligns itself with an expression of power. But the way we have reached such a result is by largely enforcing our logic on the issue. In the twenty-first century, if we see vast amounts of money being spent on something with little practical value we are often slightly repulsed. Nonetheless, in the case of aqueducts such a feeling would be born out of our rationalisation of the actual water. The water of the modern world is often silent, with meaning imposed upon it and its spatial agency greatly restricted; therefore it can be seen as a strict commodity or resource. As a result, we are far more inclined to see the water in Roman aqueducts or bathhouses as an element concerned with an economic rationale of requirement and surplus. The idea of 'waste' can only be achieved if you impose defined urban requirements on the incoming commodity. If the water within these connected urban systems bore a varied and complex meaning, it becomes significantly more difficult to say that it was being wasted in places like Rome. Perhaps the most obvious marker of a difference in the conception of Roman water waste is the fact that the sewer system was religiously profound.

The Cloaca Maxima is discussed extensively in classical sources, and even manhole covers like the Bocca della Verità bear this religious connection. As archaeologists have discovered in the analysis of other eras and locations (see Hill, 1995), the issue of waste is not easily translated into the past. This changes the equation we are dealing with when trying to estimate the value and purpose of constructing and maintaining such systems. Emphasising the agency of the water makes us think more about the potential of seeing urban water networks as part of wider natural contexts, and with this change in approach comes the opportunity to see broader possibilities for display of power and local beliefs.

However, this complexity is not often witnessed in the analysis of Romano-British evidence. When Wacher (1975, 1995), for instance, uses the word 'water' it almost exclusively refers to a practical resource that is required to supply the buildings of the town or for a tactical advantage. The water structures of his towns are a civilised aspect of a distinct incoming 'Roman' organisation of urban space. Moreover, the 20 years since the second edition of Wacher's widely read work have not been witness to a radical change in this presentation. The vast majority of excavation reports dealing with urban water features of the Roman period stick to an accepted notion of function – something that rarely considers the potentially complex nature of associations with the medium of water.

The recent discussion of the Winchester aqueduct has a complete absence of information on the function/reason for the feature beyond generic issues of practical supply (see Ford and Teague, 2011). This is despite the fact that it supplied an area which would later be satisfactorily provided for via well water and the conduit also appeared to be aligned to key aspects of the Iron Age occupation of the site. Similar sentiments could be expressed about the work analysing the Dorchester aqueduct (see Putnam, 1997). Examples of wells (see Blair, 2002) and bathhouses (see White, 1999) presented in a correspondingly familiar fashion are explored throughout this book.

This outlook is so ingrained that it is even found in recent scholarly contributions that have consciously looked to avoid traditional discourse. Mattingly (2007) has advanced a view of Roman Britain that actively rejects issues of Romanisation. Yet his presentation of aqueducts and bathhouses as 'amenities' representative of an overall identity package of Roman urbanism (ibid.: 280) is unnervingly similar to Wacher's (1975) account. Indeed, baths are portrayed as 'commonplace', with any differences merely 'exceptions' to a general rule (Mattingly, 2007: 283). Similarly, aqueducts are defined by their role as part of an organised civic service (presumably including wells and sewers) that is a 'good index of the relative success of urbanism' (ibid.: 284).

The recent work of Adam Rogers (2013) has clearly made the argument for waterscapes in Roman Britain being profound sources of symbolic and ritual significance. His focus is primarily on issues of land reclamation and adjustments to what one might call 'natural' water interfaces – e.g. drainage of fenland, diversions of rivers, waterfront installations. However, Rogers notably avoids sustained discussion of Roman water supply and the clearest architectural beneficiaries of these systems, like bathhouses. In his brief summary of this evidence, Rogers makes it clear that structures like aqueducts, sewers and bathhouses should be interpreted in light of pre-existing beliefs and associations (ibid.: 138). He also warns of the dangers of applying our modern logic to these features (ibid.: 133). Yet it is interesting that, even in this progressive account of Roman Britain, these elements of urban water are categorised as 'artificial contexts' (ibid.: 129). As a result, there is still a sense of separation of urban water networks from their natural context, something that closely aligns with our modern worldview.

Modern water supply and the urban setting

We are guided into such dialogues by the aforementioned emphasis on equating Roman and modern urban centres. Today, the provision of water to cities is an issue of paramount importance. Moreover, is there a more pivotal term than 'running water' in constructing our definitions of civilised urban living? Through the work of government initiatives, the media and charitable organisations we are keenly aware of less prosperous areas of the world, where the contrasting lack of water supply creates a situation in which disease, poverty and danger abound. These campaigns often emphasise 'simple' additions of features like water pumps that can have transformative effects for communities in these locales. There is undoubtedly an element of truth to the power of water to change lives, but this is not a rapid process. In this scenario we are often shown natural water as muddy and disease-ridden; in contrast, the output from newly installed pipes/pumps is seen as glistening 'good' water that is bountiful and pure. There is no sense of where this new water came from, or how the pump was constructed; it is all presented to us in a rather miraculous fashion.

While such campaigns have good intentions, they also serve to highlight some rather stark trends in Western thinking in regard to water. The first is that 'urban water' is abstracted from its meaning in nature, with both positive and negative connotations. The second is that because of the lack of relationship to the means of water production, we devalue the agency of the local population living in these areas. By inferring that the water is easily found, we question the independent ingenuity and organisation of the community in question. The result is a strong example of how thoughts regarding water continue to have a robust colonial tone, and it has clear connections to the interpretation of Roman-period evidence from a peripheral province like Britain.

This perspective with which we approach the provision of water for communities in less prosperous areas of the world is just a microcosm of the Western relationship between water and the modern city. There is more water running through our settlements than at any other point in history, yet we are less aware of this than ever before. In Britain, despite regular rainfall throughout the year, only the most inclement weather leaves any long-lasting effect on the streets. Our water is, by and large, hidden beneath the ground, out of sight. This process is a deliberate effort to make our lives safer and more efficient. When we require water we merely turn the tap on and are provided with as much as required; when it rains, we can usually be safe in the knowledge that a flood will not stop us in our daily routine. We live in settlements where a mastery of water has mostly removed its presence from the local landscape.

However, perhaps more troubling than the aesthetic consequences of this change is that we have become mentally disconnected from water. Kaika and Swyngedouw (2000) characterise this as being part of a Marxist development of 'fetishisation'. They outline how we have become detached from the labour and social relations involved in the process of water procurement (ibid.; Kaika, 2005): we do not know how it is transported, where it originates or the processes to which it has been subjected.

Modernity has silenced water, transforming it into a homogeneous substance known as H_2O (Kaika and Swyngedouw, 2000), thus its meaning is often dictated by the building or room in which it appears. In a domestic setting, for example, this might be why many people feel uncomfortable drinking water from the bathroom tap instead of the kitchen tap, despite there usually being no difference in the quality of water.

Part of this change is undoubtedly due to scientific advances in the arena of health, which have deemed river water unhygienic for drinking and, indeed, interaction in general. In London at the time of the Enlightenment there was an awareness that the Thames had become polluted, and efforts were made to alleviate the situation (Ackroyd, 2008: 273). In the nineteenth century this led to many notable rivers being forced underground; the Walbrook and the Fleet are prominent examples (Myers, 2011). As time passes the water of cities and towns has become predominantly hidden. Indeed, Laurence (1994a) notes how late Victorian and Edwardian views of the city targeted environmental conditions as the source of socially unacceptable behaviour, rather than poverty. Moreover, this was part of a wider Northern European movement that emphasised the rigorous control and domination of water as being an essential part of creating prosperous and economically productive nations.

David Blackbourn's (2007) book *The Conquest of Nature*, for instance, outlines how essential the control of water was to the creation of German national identity, with a particular focus on the reclamation and canalisation programmes of the 1800s. 'Canal enthusiasm' was in this period a hallmark of many prominent German writers, who wove a narrative of progress and moral improvement into the act of water management. Blackbourn (ibid.: 142) notes that writers like Ludwig Starklof in the mid-nineteenth century were at pains to emphasise the clear difference between communities in Oldenburg and East Friesland, based almost entirely upon the fact that the latter had an advanced canalisation programme and had therefore created more commercial opportunities. In this perspective, Oldenburg was backward and chaotic but East Friesland was a place defined by growth, advancement and social progression. Removing water from the landscape was a key target for influential rulers like Frederick II of Prussia. The vast marshes of Germany were seen as a dangerous dwelling for unwanted predators, like bears and wolves (ibid.: 50). By removing the water and marginalising nature, one could create productive and modern human landscapes.

It is through such projects that we are left with the pervading relationship between urban centres and their local waterscape of the twentieth century. While rivers still run through many settlements, they are seen at best as a marginal influence on urban life. Indeed, asking the general public about the main features of the town they live in will often not elicit any mention of rivers or natural water features. In essence, the urban water of today is found within buildings, wholly detached from its point of origin and natural meaning.

After such reflection, it is clear that our current relationship with water in Western Europe is primarily a product of modernity itself. Within this setting, the flow of water into the urban arena is a hidden process that is given meaning by the (human-made) space it enters. There is an inherent fear and insecurity present when this situation is altered. Because our relationship with nature is one characterised by dominance, any real display of nature's power produces a negative feeling within the modern urban dweller (Kaika, 2005: 65). A burst pipe within the house, for instance, creates a situation whereby one becomes brutally aware of the complexities and realities of the water flow supplying the building. A flash flood within a town reminds us of a precarious geographical position; a riverside view can be an advantageous element of a residence, but the reality of standing in the floodplain of a watercourse is often blissfully ignored. These dramatic events can temporarily shatter the identity of the urban space and create a feeling of unease. Kaika (ibid.) has labelled this feeling the 'urban uncanny'; in the context of water, this is when a temporary resonance of natural power alters the familiar sense of place. It is the result of viewing nature as an obstacle that hinders human development and progress (Kaika and Swyngedouw, 2000: 126). But we have to understand that these are sentiments forged over the last couple of centuries; it does not necessarily follow that these attitudes are analogous to the human perspective 2,000 years ago.

Nonetheless, accounts of the Roman period continue to feature this crossover between modernity and antiquity. It is dangerous to assume such parallels, and this is one of the critical points that this book hopes to address. Throughout history the perception of nature has been far from consistent: in reality there have been many different and 'contested natures' (Macnaghten and Urry, 1998). Water is a vital part of this, as a substance that has required consistent human engagement and entanglement. An emerging trend in scholarly and scientific works of the last ten years has been the suggestion that we have entered a very different relationship with nature in the early twenty-first century, with the pressures of climate change becoming more evident to the general public. In fact, it is likely that we need to take action to modify this relationship further, whether that be through initiatives like geoengineering (Morton, 2016) or sustainable 'rewilding' (Monbiot, 2014). These studies highlight the adaptability of the human relationship to nature historically, and the potential future approaches we should take. Water is, of course, a key part of these discussions, as a factor in the increasing flood and drought events the world will likely experience moving forward.

Over the next 100 years it will be vital that we change the relationship between urban centres and water, moving away from familiar dialogues of dominance and developing a more thorough understanding of water provenance and potential impacts on human living space. On a geopolitical scale, climate change will place water at the forefront of interactions between wealthy and more impoverished nations (with the latter likely to bear the greater burden of catastrophic changes). Relying on an outdated sense of colonial relationships, where problems can be solved by the gift of technological innovations overcoming natural obstacles, will not be appropriate or sufficient. A discussion of this situation is relevant for this book because, as outlined in the previous section, the Roman period is, even today, used as a justification for the maintenance of the current status quo. That school of thought would suggest that we continue to seek clear dominance over natural resources because it is the logical continuation of thousands of years of progression and advancement. The intention of this book is to outline how successful interactions between water and urbanisation in the Roman period, even in the context of an expanding imperial power, were actually highly complex and driven by a myriad of local factors and associations. By diversifying our view of past cities, it is perhaps possible to open up a dialogue on the need for diverse and local approaches in our urban centres of the future.

The strange water of prehistoric temperate Europe

The existence of a far more diverse and strange relationship with water has been a long-held consensus among scholars working on the prehistory of temperate Europe. This is primarily due to the frequent deposits of material culture found in direct association with springs, rivers, lakes, marshes and islands (Fitzpatrick, 1984; Aldhouse-Green, 1986; Bradley, 1990; Webster, 1995; Willis, 1999; Hingley, 2006; Yates and Bradley, 2010). Such activity is characteristic of sites from the Bronze Age through to the Late Iron Age. As the appearance of items in the archaeological record at these points can often not be satisfactorily explained in a purely practical sense, archaeologists have looked to symbolism and ritual as key aspects that help explain the relationship people had to these places. The fundamental link between water and strange 'non-practical' behaviour is so entrenched within the interpretative framework of prehistorians that even some 30 years ago Fitzpatrick (1984: 179) described it as 'unsurprising'. Moreover, this tradition of significance is not isolated to Britain – it is largely agreed that throughout prehistory people all over temperate Europe shared a similar attitude in respect to the special nature of watery landscapes (Bradley, 1990; Coles, 2001; Larsson, 2001; Stjernquist et al., 1997). The breadth of this activity is shown by the fact that it has become a supporting pillar in the idea of a pan-'Celtic' identity in the Iron Age (Aldhouse-Green, 1986), even if this term reflects a rather simplistic generalisation of societies of the time.

However, while it is fair to say that water remained meaning-laden throughout the period from the Bronze Age to the Late Iron Age, the specific ways that people acknowledged its power did vary over time. This is to be expected, as communication and engagement with the supernatural, even in a more structured and monotheistic religion such as Christianity, have been characterised by a great deal of variability throughout history. The deposit of weaponry into watery places is perhaps the most discussed sign of the importance of such locations in prehistory, and it predominates in the Bronze Age evidence (Bradley, 1990). As Fitzpatrick (1984: 179) remarked, the frequency and richness of such deposits meant that they became the beginning and end of the discussion of the significance of these areas. Unfortunately, this created a set of circumstances whereby archaeologists projected continuity between the Bronze Age and Iron Age, and did not adequately explore how interaction with water was changing. Sites such as La Tène in Switzerland, with its rich metalwork (particularly weaponry) finds in watery contexts, seemed to echo similar values as in the Bronze Age, albeit with the type of metal changing. Yet as Bradley (1990) noted, this obscured many ways in which the evidence of the Late Iron Age differed from that of earlier periods.

Even at La Tène there was a breadth of items recovered beyond weaponry, with coins and currency bars two of the more prominent new find types. Bradley (ibid.: 173) notes how these inclusions are reflected in other watery contexts, even in Britain, during the Late Iron Age. There is also the fact that food-related items became increasingly involved in ritual during the Late Iron Age, perhaps to signify elements of fertility and prosperity. Fitzpatrick (1984: 179) mentions how such organic votive deposits were unlikely to have been recovered from watery places, and yet could have played an increasingly prominent role during this period. His analysis of finds of the La Tène era from the River Thames uncovered the possibly votive deposition of items like bronze cauldrons, tankards, bowls and wood sculptures (ibid.: 166). This increased diversity of find types above just weaponry could well reflect an escalation in the interaction with water for more varied ritual and symbolic reasons. The possibility of individuals, or groups, using food in votive offerings also implies a degree of ephemeral activity that is difficult to assess from an archaeological point of view.

Another development highlighted by Bradley (1990) was the increasing structural interaction with waterscapes in the Iron Age. In previous eras there is much evidence suggesting a link between water and burial. In locations as diverse as Orkney (Scotland) and Scania (Sweden), archaeologists have made explicit links between burial mounds and their surrounding waterscapes (e.g. Bradley, 2000; Phillips, 2004). In Britain there are a number of examples of Bronze Age burial activity close to rivers; the Thames, Witham, Tas and Yare are all mentioned in subsequent chapters in this regard. This connection may also be substantiated by the apparent similarity in the deposits found within burial context and those found in watery contexts (Bradley, 1990: 103).

With the diversification of items found in water during the Iron Age comes a development in the type of features regularly associated with these areas. The site mentioned earlier at La Tène is a prime example of this, with the majority of votive deposits being found between two large timber bridge/platform features. They are not isolated instances, and similar features can be found at sites in Britain such as Fiskerton, where a monumental causeway was the focal point for votive activity (Field and Parker-Pearson, 2003); the Fens area of East Anglia saw precedents in the late Bronze Age, and at sites like Flag Fen such activity continued through to the Iron Age (Pryor, 1992).

The Late Iron Age also witnessed the emergence of shrine features that coincided with liminal watery areas in the landscape (Bradley, 1990: 175; Webster, 1995). Some of these were more defined ritual enclosures, akin to the continental *Viereckschanzen*, which contained deep shafts with ritual offerings. The depth of these shafts may indicate a direct link to underground water, or at least they could have contained rainfall. In fact, as Fitzpatrick (1984: 182) notes, the frequency of such shrines could have been underestimated, as they would not always be readily detectable from an archaeological point of view. While votive deposition of objects still forms a crucial part of the activity related to waterscapes, the increased prevalence of built structures provides us with a sign of interaction that does not necessarily need a particular deposit to make it meaningful. Other monumental interactions increasingly associated with waterscapes in this period are pit alignments and earthworks, which present us with further evidence of significant communal structural engagement with water that is not purely practical. Both feature types may have defined land organisation between communities, but their relationship with water may also have held critical symbolic and ritual messages that underpinned such practical functions (Bryant, 2007; Chadwick, 2007; Rylatt and Bevan, 2007).

In many ways these developments represent an increased hybridity in the waterscapes of Britain by the Late Iron Age. Instead of being locations more representative of another world, or portals to such worlds, and in a sense marginal in previous eras, they increasingly became a mixture of 'human-made', 'natural', practical, symbolic and ritual aspects that formed an integral part of the life of local communities. It is perhaps tempting to suggest increased human engagement leads to normalisation, or rationalisation, that would erode the special nature of these places. However, in recent years there has been an emphasis on deciphering the symbolic impact of these structural developments of the landscape/waterscape. Archaeologists have explored how they could be evidence of the new ways people were manipulating and respecting traditional sources of power (e.g. Tilley, 1994; Evans, 1997; Bryant, 2007; Rylatt and Bevan, 2007).

The consequence of embracing this tradition, in both its scholarly direction and its tangible archaeological results, is a troubling situation in respect to the interpretation of the Roman cities that came to characterise these same landscapes. The traditional response of Romanists has been to 'deal with Iron Age problems in purely Iron Age terms' (Rivet, 1958: 75), and view the Roman period as clearly distinct from such activity. Such compartmentalised views have undoubtedly affected how archaeologists have interpreted evidence. In the last 10 to 15 years, for example, there has been a call for the exploration of potential continuities from the late pre-Roman Iron Age into the subsequent period (Creighton, 2001; Millett, 2001); unusual votive depositions in prehistory have naturally been an area of focus in this regard. Indeed, Fulford's (2001) work in Silchester highlights the pattern of unusual deposits beneath many of the Roman buildings, and Collie's (2013) more recent work synthesises this evidence and compares it to similar depositional practices at St Albans and Dorchester. In addition, Merrifield and Hall (2008) note the potentially important pre-Roman religious activity in London. Yet in some ways this still results in the archaeologist isolating the Iron Age in its own terms. A concentration on votive depositions and dialogues of direct continuity from a previous era mirrors the concerns Fitzpatrick (1984) voiced 30 years ago regarding water traditions of the Bronze and Iron Ages. As the changing practices of the Late Iron Age suggest, seeking continuity is likely to present one with a limited perspective on the evidence.

In this regard, it is important to note that some of the most lauded and celebrated aspects of the Roman town are intimately involved in the management of water: the baths, aqueducts, sewers and wells are all fundamental parts of the image of everyday life in the Roman period. However, if anything these elements are seen as the diametric opposite of any Iron Age activity, even if they deal with the same medium in the same landscapes. Overall, the disparity between this widely replicated treatment of water in Roman Britain and its agreed significance in the Iron Age is readily apparent. The development of these structural interactions with water appears to nullify any meaning for the Roman archaeologist, despite the fact that its importance was undiminished through comparable changes in the Late Iron Age. This perspective is made even more puzzling by the recognition that many of the Roman towns in Britain are directly informed by previous Iron Age settlements and show signs of acknowledging previous ritual and symbolic concerns (Willis, 2007a, 2007b). Rogers (2008: 73, 2013) also rightly emphasised a direct pattern of association with water: out of the 21 Romano-British towns he analysed, 20 were found to be near water features. The majority of these settlements were built on the floodplains of nearby rivers, in areas that were often marshy and thus impractical from a modern point of view. Unfortunately, a consideration of the potential symbolic and ritual resonance one could derive from interacting with these contexts has mostly been ignored in favour of implementing a very different generalised Roman relationship to water.

With the prevailing view of prehistoric waterscapes in mind, we must question our portrayal of later periods. The relationship that people had with water during this pre-Roman period appears to be considerably different from our own. The character of these differences is explored in more detail in subsequent chapters, and it suffices to note here that these practices would be 'strange' in our modern urban context. Besides, we have evidence for these practices encompassing a chronological timeframe much longer than the subsequent Roman period. By the Late Iron Age the specifics of interaction with water may have changed, but the importance of this relationship was apparently undiminished and still not easily understood from a modern perspective. Indeed, if anything the development of more variable practices and structural interactions has created a situation where archaeologists have had to examine local contexts more thoroughly from different perspectives to understand what was happening in the past. We return to this topic of pre-Roman water interactions in Chapter 3, but it is clear that prehistorians have been far more open to the idea of 'making strange' interpretations of water in the past, moving away from the modern rationalisation we have already discussed.

Water and hybrid urban identity

As the subtitle of this book suggests, the term hybridity and its relevance for the understanding of urban identity in Roman Britain are a crucial theme throughout the subsequent chapters. It is proposed that this term provides a more accurate description of the complex entanglement of cultural change that we see in places like Roman Britain – especially in relation to the presentation of water. In its broadest sense, a hybrid represents a combination of different inputs to create a new, and possibly unique, output. Critics of the term argue that this process acknowledges defined inputs that come together to create a definitive new hybrid product. Thus there is a danger that such terminology could propagate usage of outdated labels; regarding analysis of Roman Britain, the argument is that discussing hybridity endorses continued use of 'native' and 'Roman' dichotomies that combine as part of a hybridisation movement (Webster, 2001). However, this book intends to show that the use of such language can facilitate us moving into areas of multiple identities, by highlighting how interactions with water display a complex and changing reality in urban life at the local level of Romano-British settlements.

Thus the objective is not to create a new universal theory that can be applied to any town. While broad trends are naturally highlighted, the perception of water in the past creates an acute sense of local diversity. So similar features may align, but actual meaning is still shaped by a uniquely local waterscape and the perspective of the people experiencing it. So one does not anticipate a definitive hybrid product or a consistent hybridisation process; rather, there is an expectation of showing a meaningful language of water but with notable differences in dialect and expression within the various towns of the province. Another part of that language is the incoming associations that result from Britannia becoming a province of the Empire, which is why ideas of linear continuity can be just as misleading as emphasising an alignment to a universal Roman standard.

While this introduction has given a survey of the genesis of our interpretations of water in the Roman period, and highlighted some of their shortcomings, the next chapter is dedicated to highlighting the acute local diversity inherent in the treatment of municipal water in Rome and other Mediterranean cities. While features like bathhouses and aqueducts feel like culturally consistent aspects of the Empire, even in these core areas of Roman influence they can be interpreted as complex hybrids. As noted above, in the Mediterranean we are dealing with meaning-laden water that could define urban space. Even in the ancient sources, apparently human and engineered structures of municipal water supply were not necessarily separated from the wider 'natural' waterscapes that have been widely seen as ritually and symbolically charged by classical scholars. Accordingly, features like wells, aqueducts, sewers and bathhouses were not necessarily given consistent definitions, even in Rome. In reality they were diverse expressions of local beliefs and practices, mixing human, natural, practical and religious aspects. While they had a degree of structural consistency, the medium of water created a baseline diversity that could rapidly alter their spatial meaning in an urban setting. In the Mediterranean Roman world, water was acknowledged in similarly symbolic and ritualistic fashion as in temperate Europe. It bears little resemblance to the passive and homogeneous water of our modern cities. The point of exploring these issues is to emphasise the fact that if water supply features can be 'made strange' in a Mediterranean context, it makes it even less likely that we can rely on a unified 'Roman' presentation of them in Britain, where there were strong pre-existing examples of meaning-laden local interactions with water.

Outlining this diversity of urban water features creates a situation where the reception of new interactions with water in the Roman period in Britain was likely anything but straightforward. Even bearing in mind the progressive work of archaeologists like Rogers (2013) and Willis (2007b) in recent times, the disparity of scholarly water analysis either side of the Iron Age and Roman conceptual 'divide' does not make sense in rational terms. The municipal water of the 21 'primary' towns of the province, as outlined by authors such as Wacher (1975), is still predominantly seen as being constructed in alignment with a culturally uniform Roman identity package. The suggestion that in reality the incoming Roman tradition of water management could have been very open to incorporation of local associations makes a reconsideration of the evidence even more pressing. It is therefore necessary to look in detail at a selection of these towns and make an effort to understand the pre-Roman context of water interactions, ascertaining the extent to which we can see deliberate interactions with these in the Roman period. Bearing in mind what has been said above, it is likely that in these centres there will be clear elements of crossover between the Iron Age and Roman-period engagements with water. Of course, the precise purpose of these actions should also be a product of local circumstances, and we consequently should be open to accepting, for instance, discrepant experiences of buildings like a Roman bathhouse. Exploring the unique characteristics of these prominent Roman waterscapes in Britain can give value to features that have been unfairly judged against their continental contemporaries, largely on the basis of scale. Moreover, we can see how even a subject as traditional as water management can play a role in the progressive discussions about Roman identity currently appearing.

Ultimately, the water features discussed in this book are traditionally identified with a civilised Roman way of living that we continue to underline as an inspiration for modern urbanism. Traditional approaches imply that Roman engagements with water, while not entirely analogous, essentially provided a foundational example of the logic that would come to define twentieth-century urban water management. It is a sense of familiarity that has served Roman archaeologists and historians well, easily creating relevance for their work with a wider audience. However, 'making strange' or defamiliarising aspects of Roman urbanism, as is the intention in this book, should not result in a lack of relevance. On the contrary, instead of merely being the bulwark of an argument that sees our urban status quo as being historically justified, it may create a situation where we can more clearly see the possibilities of proposed future directions. In an effort to create sustainable settlements, the recent work of writers such as Edgeworth (2011), Montbiot (2014) and Sedlak (2015) outlines a future of urbanism that is keenly focused on the local connections between natural environment and human development. These are worlds defined by a keen sense of hybridity at the local level, understanding the overlapping relationship, for instance, between water consumption and the sources from which water can be obtained in the urban landscape. It is therefore one of the intentions of this book to propose that engagement with water in Romano-British towns was fundamentally different to that in the cities of today. Yet the differences outlined here can actually be helpful in accepting the possibility and necessity of change in our future approaches to urbanism.

Chapter 2

Hybridity in classical accounts of urban water

Meaning-laden Roman water

The interpretive approach to water in Roman Britain, outlined in the previous chapter, is so troubling because it is not even a fair reflection of the evidence from the capital city of Rome. Rivers and springs are the primary sources of water flow in the Roman town, so assessing their treatment in the classical context is vital. The city of Rome remains intimately connected to the flow of the River Tiber. Its characteristic twists and turns frame some of the most important ancient sites, such as the Campus Martius and the Forum Romanum. While we are accustomed to picturing Rome on seven hills, the administrative/religious centre of the city was deep within the thralls of the Tiber's floodplain. The powerful nature of the river meant that regular flooding was a near certainty in the Forum (Aldrete, 2007); this is something which still causes problems to this day (Ammerman, 1990: 637). As a result of this placement, the perception of these great monuments would have been inextricably linked to the Tiber itself – more than just a feature, it was an entity that defined space.

The anthropomorphised deity of the river, Tiberinus, is particularly prominent within the historical foundation myths of the city. Plutarch, for instance, notes that when the infant founders of Rome, Romulus and Remus, were placed on the edge of the Tiber, it rose up, gently took them downstream, and safely deposited them near to a fig tree (Plutarch, *Vitae Romulus*: 3–4). After this intervention the infants are discovered by the She-Wolf and set on the path to greatness. Similarly, in the *Aeneid* the Tiber comes to the protagonist Aeneas in a dream and guides him towards Latium, and the eventual site of Rome (Virgil, *Aeneid*: 8.26–103). Again, this intervention occurs at a pivotal point in the narrative and, in this case, gives Aeneas a distinct focus on his eventual goal. In addition, whenever the classical sources mention the river it is in the autocratic language one would expect to be reserved for a king, for example 'the ruler of waters' (ibid.: 8.77) and 'most eminent of all' (Ennius, *Annales*: 66–69). The

Republican and early Imperial Roman custom of avoiding the terminology of sole rulers makes this treatment somewhat conspicuous.

Unsurprisingly, Rome's religious tradition was also profoundly entangled with the waters of the great river. Many of the religious practices that were followed in the city through the Republican and Imperial periods had a shadowy origin in the Age of Kings. Numa Pompilius was the second King of Rome, a semi-mythical individual who is attributed as the central creator of such traditions (Hooker, 1963: 91). Livy tells us that he proclaimed many of his reforms to have been a consequence of consort with the water nymph Egeria (Livy, *Ab Urbe Condita*: 1.19–21). Moreover, Numa is said to have created the office of the high priest Pontifex Maximus, which translates directly as 'Greatest Bridge Builder'. Holland (1961) convincingly made the argument that this title was the crystallisation of the early Roman importance of streams and waterways. Back in this archaic period the topography of the city was defined by a proliferation of such features, which could only be crossed by performing the religious rite of the *auspicia perennia*. As such, the establishment of permanent blessed bridges was an act deeply bound to the religious beliefs of the city. Even the Vestal Virgins (also founded by Numa) seem to have a definite link to this watery tradition: the ceremonies associated with the Pons Sublicius involved the Vestals throwing effigies into the Tiber from the bridge (Dionysius Halicarnassensis, *Antiquitatum Romanarum*: 1.38.3; Ovid, *Fasti*: 5.621–622), a practice that could have started as human sacrifices, and also has mythic connections to the creation of the famous Tiber Island. These connections to bridging water appear to be an early example of how establishing human-made additions to a waterscape, even if practical reasons were clearly prominent, could enhance the religious associations already present.

However, while the Tiber inevitably holds a special place in the mythic origins of Rome, other rivers were also perceived as far more than simple elements of geography. In one of his letters, Pliny the Younger (*Epistulae*: 8.8) describes the Italian River Clitumnus (the Clitunno in Umbria) in a passage that strains towards the metaphoric (Murphy, 2004: 139). The mention of shadowy groves and offerings of coins suggests a type of reverence. This is underlined by the personification of the water as the god Clitumnus, who is described as clad in a purple-bordered toga (a clear sign of authority). Just like the Tiber, these other rivers of Italy play active roles within the lore of ancient writing. Ovid (*Amores*: 3.6), for instance, depicts the River Anio consoling the character of Ilia, offering her refuge within his kingdom. Ilia is a pivotal individual, being the mother of Romulus and Remus but also the descendant of Aeneas. Moreover, this emphasis on the importance of watercourses ranges beyond the limits of Italy. In his *Natural History*, Pliny the Elder (*Historia Naturalis*: 3) depicts the rivers of Northern Europe as the key determining factors in the formulation of identity and sense of place (Murphy, 2004).

There is a sense that, because of its deep-rooted cultural and historical value to the Roman people, water could have presented a medium of legitimisation for new rulers and their respective foundations. A vivid example of this is the plans of Julius Caesar, who after gaining power sought ways to solidify his support. His chief proposal was the invasion of Parthia (avenging the defeat of Crassus), but he also planned to shift the course of the Tiber to create a new central area, perhaps intended to supersede the Campus Martius (Purcell, 1996). Similarly, when Trajan conquered Dacia he made a conscious effort to incorporate a personification of the Danube on his coins and monuments. The reclining pose of the personified Danube is similar to depictions of Tiberinus, further emphasising the importance that Romans attributed to provincial rivers. Nor is this just synonymous with Trajan: the numismatic presentation of rivers is almost exclusively in these personified terms, with coins from different eras and locations embracing this symbolism (Figure 2.1). This serves to underline the powerful individual divinities that rivers possessed in the Roman world, quite apart from the homogeneous value often given to them in modern times.[1]

This is significant, because while the evidence garnered from the literature is valuable, there are always going to be doubts about the extent to which writers of epic poetry are reliable sources of broader contemporary views. The numismatic evidence seems to legitimise this as imagery designed to be recognised by the common people. Indeed, it not only shows us that the water of rivers had a religious aspect, but also exhibits what a valuable ideological tool this provided. On both his coinage and his eponymous victory column, Trajan chooses to frame his military success in terms of controlling the Danube. Such conscientious replication of an image suggests the Roman people would have immediately understood the power that is implied. This conception of rivers appeared to be alive and well within the High Imperial period and was being utilised by one of the most successful emperors in Rome's history.

While perhaps the idea of divine rivers seems rather alien to us today, the importance of springs feels slightly more familiar. No doubt this is principally due to the continued recognition of these sites by later Christian activity (Jones, 1992; Quinn, 1999; Davies and Robb, 2002; Sauer, 2011). The strength of the Roman tradition is well attested in the ancient literature. Servius, for instance, famously proclaimed 'no spring is not sacred' (*Nullus enim fons non sacer*) (Servius, *In Vergilii Aeneidem*: 7.84). This was in reference to the much earlier comments of Virgil, showing an enduring reverence for these locales; something that is illustrated vividly in Rome, where many springs were sites of worship. Debatably the most important was the spring of Juturna, located adjacent to the Forum. Again, this watery feature is woven into the fabric of Roman history, apparently being the place where Castor and Pollux watered their horses after the

Figure 2.1 Examples of reclining river gods on Roman coins. (A) Von Aulock, *Lykaoniens* 85–88 reverse – river god reclining left; (B) RIC 440 reverse – river god Tiber reclining left; (C) ACANS 12A04 reverse – river god Strymon reclining left.

Sources: (A) image courtesy of www.cngcoins.com; (B) image courtesy of Australian Centre for Numismatic Studies and Museum of Ancient Cultures at Macquarie University; (C) image courtesy of Australian Centre for Numismatic Studies.

battle of Lake Regillus (Plutarch, *Vitae Coriolanus*: 3.4). Furthermore, it is known to be the preferred source for water rites involving the Vestal Virgins: supposedly they collected water from the spring in specially shaped vessels designed so the carrier could not rest them on the ground (Wildfang, 2006: 11); this was to prevent any spillage, further underlying the importance of what they held. Rome was full of other springs, and all had different deities that were said to reside within their locality. The aforementioned nymph Egeria, for example, had a spring located near to the Porta Capena that has survived to this day. This is the place where Numa is said to have consorted with her, and thus it is tightly woven into the religious narrative of Rome.

There are similar prominent examples of famous springs throughout the Empire, and it seems that such locations played a significant role in the development of Roman towns. Some settlements gravitated entirely towards their local spring; towns like Aquae Sulis (Bath, England), Aquae Segetae (Sceaux-du-Gatinais, France) and Aquae Statiellae (Acqui Terme, Italy) even show it in their names. They remain famous for their thermal springs to this day (Hodge, 1992: 263). No doubt they developed this importance in part due to the interesting and diverse qualities of the water, as well as its origin from deep within the earth – an area that represented another world (Bradley, 2000). The common people's perception of springs would undoubtedly have erred towards the supernatural, with layers of folklore adding to this mythos. Even a learned individual like Vitruvius, who analyses these features in relatively scientific terms, reveals to us a world where springs can turn people foolish, make all their teeth fall out, make them abstemious or, indeed, give them better singing voices (Vitruvius, *De architectura*: 8.3.22–24). These descriptions would have served to intertwine the springs of the Empire further into the mysterious world of dryads and nymphs.

However, as noted in Chapter 1, it is clear that these aspects of Roman belief have had relatively little impact on our view of ancient municipal water structures. Much of this relies on the very modern idea that there is a definitive separation between the human and natural worlds, and the former has a corruptive effect on the latter. In this school of thought, springs and rivers can have far-reaching significance, but as soon as they are brought into the human/urban sphere this special nature is eroded. This idea was championed by Wissowa (1912: 180), who proposed that the 'living water' of the Roman natural world was prized, but its significance would have been lost once it interacted with the urban landscape. It is an approach that feels very in tune with the modern world, where we emphasise the purity of abstract natural water sources, placing them above our own local piped water supply. The unprecedented market for bottled water in modern Britain is evidence of an exploitation of these attitudes, relying on a perceived natural purity of such water and using it as a point of separation from the piped supply already provided to homes. However, we have to recognise that applying this way of thinking to our historical evidence is highly contentious.

The classical sources are often seen as endorsing this familiar rationalisation of water, with authors like Frontinus and Vitruvius cast as the direct forebears of our modern engineers and architects, putting their emphasis on the practicalities of supply systems rather than the more abstract belief systems of which they are a part. The principal aim of this chapter is to challenge this idea that Roman urban water needs to be considered in abstract from the meaning-laden Roman waterscapes outlined above. By consulting the work of classical writers, we can see how urban water networks were viewed in hybrid terms, even in Rome itself. As a result we can envisage how bringing consistent water to towns and cities in the Mediterranean was an act of entanglement that bound local natural contexts to human architecture. Accordingly, it could represent both the pragmatic and the significant symbolic/ritual associations of local communities. This chapter discusses how this water at its source, in its journey to settlements and in its presentation in key central structures of the Roman city embraced hybridity at the local level.

An entangled source

It stands to reason that the best place to start when re-evaluating Roman water networks is their sources. As outlined previously, springs and rivers were ritually and symbolically charged parts of the Roman world. They represented points of interface with a diverse, personified and acutely powerful nature. However, the fact that we have so much writing from classical sources on their varied importance suggests that they were profoundly connected to various human activities. After all, any spring with beneficial properties is unlikely to have been left alone. As Bradley (2000) suggests, throughout temperate Europe natural places of significance became increasingly monumentalised, with changing aspects of engagement and movement. But in those regions this human involvement with the natural context did not lessen meaning. On the contrary, by becoming a hybrid mixture of human engineering and natural power, the importance of such locations or features could be greatly enhanced. It is interesting how in our consideration of Roman urban water we have readily dropped this sense of a broader connection to the landscape.

Wells and springs

The topic of wells is a good starting point for this discussion because they are a fundamental aspect of water supply, representing a first engineered solution for many settlements in the Roman world. The modern reader brings a certain amount of baggage to this subject, some of it helpful and some of it not so. Of course, most people will be aware of the idea of a 'wishing well', or quite possibly other beliefs related to the enduring religious associations with these features. Many ritualistic well traditions in Britain were appropriated by the Christian faith to stem heretical practices. Strang (2004: 88) notes how Christian festivals were frequently timed to coincide with well-visiting rituals, and churches were also built over these structures; for example, the abbey at Cerne Abbas was constructed on such a site.[2] Other wells were simply rebranded with an appropriate saint linked into the local lore, moving from 'magical' to 'miraculous' (ibid.: 89). If such a transition was challenging to achieve, there was also the possibility that wells would have been translated into the extensive narrative of Christian evil (e.g. demons and the devil).[3] The pre-Christian identification of these areas with an underworld was not necessarily negative, but it was easily woven into the fabric of Christian belief to inhibit the pagan worship of such sites (Sauer, 2011: 507). This backlog of tradition, in addition to the frequency of remarkable finds, is undoubtedly why Roman archaeologists have been open to the idea of ritual and symbolic meaning for wells, even within urban contexts. However, even with these sentiments there is a tendency to emphasise the human aspect of wells, and if they are without particular human deposits then they are seen in largely practical terms.

This last point highlights some of the negative connections the modern reader will make with wells. Their human aspect has defined them in cities. For instance, in the dense urban settlements of early modern Europe wells became breeding grounds for diseases like cholera (Sedlak, 2015: 30–32). They amplified the unhygienic activities of the humans using them. As a result, today it is unlikely that the modern consumer would choose to drink bottled well water, but spring water is welcomed into many homes because of its 'natural purity'. But, as Smith (1922) noted, the terms 'well' and 'spring' are not easily separated in any language, and this is very much the case in ancient Rome. After all, a well is usually perceived as a structure that taps an underground aquifer; while the spring represents a natural occurrence of the aquifer at the surface. Conceptually, there seems very little reason to differentiate the value of the water when a well and a spring are located nearby. While some may debate the spring is a natural phenomenon, and therefore could have held more significance for those in the past, the reality is that they would also often have been given a structural framework. In the Roman period this could have turned a spring into the source for an aqueduct. In an urban context the spring would have become every bit as much an example of a hybrid phenomenon as a well. Our modern way of seeing them, positively or negatively, as human features outside nature does not necessarily hold in antiquity.

In fact, the classical sources can show us a far more complex picture. For example, in his description of how one locates a suitable site for a well, Vitruvius highlights the almost ritualistic aspect of the task. One of his recommendations involves the placing of oiled bronze bowls in trial pits and covering them in reeds to draw out the moisture (Vitruvius, *De archi-tectura*: 8.1.4). Perhaps more significantly, he also outlines how when a well taps a spring many shafts are to be dug and connected underground with tunnels (ibid.: 8.1.6). So in this case we can see how the 'human-made' well can increase the flow and yield of a 'natural' spring.

Vitruvius is not alone in giving us such a description: Seneca mentions the sinking of wells on a number of occasions. He first disproves the idea that rainwater could account for the appearance of underground supplies. He describes how in a dry area a well could be sunk to a depth of 60 or 90 metres and still find 'living water' (Seneca, *Quaestiones naturales*: 3.7.3). The use of this term is interesting, because it is often deployed in the description of springs and water concerned with ritual in the Roman belief system. The second mention occurs when Seneca relates how 'the wells and lakes disappeared because the land ceased to be tilled' on the island of Crete (ibid.: 3.11.4). Again, we are given an example where wells are paired with lakes, implying that they are all part of a natural world that is reacting to human processes. He continues by noting how 'some wells are full for six hours and dry six alternately'; the explanation, according to Seneca, being the similar sources of these wells and rivers, which run dry at some points during the year. He also makes reference to certain wells that 'throw up not merely mud but also leaves, and bits of crockery and any other filthy things that have accumulated in them' (ibid.: 3.26.6). This implies that these features had the power to cleanse themselves and reject unwanted materials. The purifying attribute could be a reason why it was deemed appropriate to sprinkle well water on food before sale in Rome (Edlund-Berry, 2006). Not only does Seneca link wells directly to other elements of 'natural' or 'living' water, but he also distinctly confirms a purifying aspect that engenders a sense of importance similar to that accorded to a spring. This purity of well water is mirrored in the direct references Pliny makes on the subject:

> From what source then shall we obtain the most commendable water? From wells surely, as I see they are generally used in towns, but they should be those that the water of which, by frequent withdrawals, is kept in constant motion.
>
> (Pliny the Elder, *Historia Naturalis*: 31.23)

Thus the idea that well water has commendable purity is emphasised, but also the fact that it is living, moving water like that of a spring. The other chief aspect communicated in these accounts is a sense of wells being subject to that same personified nature which has been mentioned previously. There appeared to be various occasions where a certain action was needed, otherwise a subsequent reaction would be seen in the well. So if the fields were not tilled appropriately, the wells and lakes would disappear, or if water sources were subject to dumping of extensive refuse they would reject this material through the wells. Even Seneca's account of wells being full and dry alternately for six hours conjures images of mischievous water deities. These accounts indicate that the act of well construction may have involved a degree of negotiation and balance with the natural world if one was to secure an enduring water source. They reflect the possibility that any failure was probably not seen in purely practical terms, but also as a product of not satisfying or placating the local deities responsible for the water. This suggests the possibility that many unsuccessful wells in a specific area may have adversely affected the perception of the space in ritual terms (or the opposite positive effect of successful wells).

The most obvious acknowledgement of well worship within the Mediterranean tradition was probably the Fontinalia festival in the Roman religious calendar (Varro, *De lingua Latina*: 6.12). Held on 13 October, this is often thought of as a festival associated with springs; yet it also involved wells being dressed with garlands and objects thrown into the shafts (Scullard, 1981). Other festivals such as the Furrinalia, Neptunalia, Camenae, Juturna and perhaps Carmenta could also have been involved in the appreciation of underground water sources (Edlund-Berry, 2006: 168). This gives us a general sense of the widespread respect that was entrenched by a series of annual celebrations in Rome and beyond. The festivities around the wells would perhaps be framed by nearby buildings (possibly by procession into central areas like the Forum); these may even have been temporarily altered to befit rituals of the particular festival. Furthermore, the materials needed to dress the well, or deposit within it, would have to be made (or purchased) somewhere, possibly in the immediate community. If so, these items may have had a local flavour to them, possibly being manufactured to suit the attributes of a particular well. Certainly, the still-existing well-dressing traditions of Britain appear to project such concerns and achievements of the immediate populace, rather than being consistent across all wells and in every year.

The watery nature of the Forum in Rome itself was outlined previously, and one of the most mysterious monuments within this space is the Lacus Curtius. The name implies a lake, and this is probably a reference to the earlier prevailing conditions in the area during the archaic period. The actual structure that developed, and has been preserved to this day, resembles a pit or well; the latter seems more fitting given the name and geology of the area. Little is known of the real purpose of the Lacus Curtius, but it seems to have maintained a ritual importance well into the Imperial period. In fact, Suetonius (*Vitae Augustus*: 57) tells us that to show appreciation for Augustus, 'all conditions of men ... each year threw a small coin into the Lacus Curtius'. This may hint at a veneration of pits/wells within the central Roman tradition. Just as importantly, it shows how in the Imperial period such a traditional ritual focal point became an ideological tool.

Wells in the Roman tradition thus provide us with a solid example of how the addition of human-made structure to water sources did not necessarily erode meaning, as is the case with modern urbanism. They represented a human entanglement with a particular waterscape, and in many ways could heighten the significance and relevance of such natural foci for people in the Roman era. While human deposits in wells will always be a source of interest for archaeologists, they are just one aspect of acknowledgement of the importance of the water. What this suggests is that we should be open to the idea that wells, like the springs of the Roman world, could often have possessed meaning without leaving characteristic objects in the archaeological record.

Sourcing an aqueduct

This brings us to the subject of aqueducts: undoubtedly the most striking and celebrated water supply structure of the Roman period, and a key focus of this chapter. In provincial territories like Britain, highly structured piped-water supply is a key point of difference in the use of water between the Roman period and previous prehistoric occupation, and this has been used to delineate very different aims. However, while the structure of aqueducts was unquestionably novel, the water sources which they harnessed are likely to have been entrenched in local culture and landscape history. The above discussion of wells can be a platform upon which we alter our perception of the link between source and conduit. If a highly engineered structure like a well can be deeply connected to wider waterscapes, there is little reason to suggest that aqueducts should deserve different treatment.

These observations are given emphasis when we start to analyse some of the sources of Rome's aqueducts. As discussed previously, Frontinus has been used by modern readers to justify a very structural and engineering-based approach to these features. His *De aquaeductu urbis Romae* has become a focal point for any study of the subject. Frontinus was an engineer of sorts, appointed to maintain the aqueducts of Rome, so writers have seen modern technical analysis as a natural extension of his treatise. The text is known as a *commetarii*, which implies a set of notes or records that could either have been for wide consumption or possibly more personal consultation (Rodgers, 2003). Traditionally writers have conceived this text as a dry, systematic account of water supply in Rome (Goodyear, 1983: 672; Hodge, 1992: 16); many have looked to it as a sort of instruction manual that could have been consulted by people assuming the same role as Frontinus. Yet as Bruun (2012: 16–18) notes, if we assess the text as an administrative handbook, it rather lacks in completeness. DeLaine (1996) has questioned whether this was the intention of Frontinus in the first place. Indeed, she put forward the notion that the text could in fact have been a political statement, possibly even delivered to the Senate as a speech. There is a possibility that the *commentarius* was, in reality, a monument to Frontinus' extraordinary political achievement of sharing a third consulship (ibid.: 136). Instead of concentrating on the specifics of this, which would naturally have to deviate to focus on Trajan (the man with whom he was sharing the consulship), the subject of the *curator aquarum* could emphasise the power of Rome and the tradition of service to the state by great men. Frontinus could be framing his achievement in this context. This would have resonated well with a Senate which, in the wake of one of its number (Nerva) becoming *princeps*, thought the power of its chamber was once again on the rise (ibid.).

This fundamentally alters the way that we approach Frontinus and his contribution to the subject. As DeLaine notes (ibid.: 139), the practical nature of the text could be much more about establishing notions of power and wonder to an audience not entirely familiar with such details. In turn, the more vivid historical sections make overt use of the physical element of water to frame Rome's ongoing legacy of power. This is shown in the way Frontinus describes each aqueduct of the city, noting semi-mythical origin stories, linking them to famous Roman figures, and then rationalising them as working structures in his own time. For instance, the Aqua Appia was one of the first aqueducts of Rome and was established at the time of the Samnite War. It shared a creator with the Appian Way, started at one of the famous properties of Lucullus and then joined with the later conduit of Augustus before reaching the city at the Porta Capena (famous for the association between the mythical King Numa Pompilius and the nymph Egeria) (Frontinus, *De aquaeductu*: 1.5). All these prominent names and places are combined with practical descriptions of distance, which bring the reality of the aqueduct into the present. Thus Rome is woven into its mythical landscape by the movement of water. With this in mind, we can re-evaluate the most famous passage of the text: the comparison between the 'indispensable aqueducts' of Rome and the 'idle Pyramids' and useless works of the Greeks (ibid.: 1.16). This has been widely regarded as a 'typically Roman' comment, deriding the more frivolous monuments of the East. A working monument like the aqueduct is far more in line with the image we have of the industrious and practical Empire. Yet maybe this is Frontinus highlighting the essential active ideological power of the aqueducts themselves. Instead of monuments such as the pyramids, which are limited to defining one area, the water of the aqueducts can be productive throughout the city of Rome. The powerful associations of the past can continue to be woven into the new creations of Frontinus' present. This interpretation rests on the recognition that productivity is not just limited to practical economic principles, especially not in the Roman period. Consequently, basing the study of aqueducts on practical supply is limiting in the extreme.

The varied associations with aqueduct water in the account of Frontinus are also displayed in the significant amount of time he spends outlining the ritual and mythic background of each of Rome's aqueducts. The waters of the Aqua Virgo, for example, were apparently named because they were discovered after a young girl pointed soldiers in the direction of the spring. A temple was duly established at the site, within which a painting documents the origin of the aqueduct (Frontinus, *De aquaeductu*: 1.9–10). This source appears to have been revered in a similar way to other springs. The painting in the temple ties the foundation of the aqueduct and the spring together; indeed, the building seems to have been erected to this end. There is no sense from the description that the aqueduct lessens the significance of the spring.

Another account from Frontinus is that of the Aqua Claudia, which he notes is sourced from the Curtian and Caerulian springs (ibid.: 1.13–14). The fact that these two springs are given specific names suggests they are of renown; the latter implies a perceived purity ('The Blue'). The strength and beauty of these springs are consequently extended to the aqueduct itself, which is described as having an almost unrivalled excellence. Suetonius (*Vitae Claudius*: 20) follows this pattern, declaring Claudius 'brought to the city on stone arches the abundant founts ... one of which is called the Caerulius and the other Curtius'. The waters of the Aqua Marcia are treated in much the same way, with Pliny the Elder (*Historia Naturalis*: 31.41) deeming it the 'glory of the city of Rome ... among other divine bounties'. Martial (*Epigrammata*: 6.22) mirrors this sentiment, proclaiming that the Marcian waters 'shine so brilliantly, and are so pure, that you scarcely suspect any water to be there'. This all comes together to suggest that these 'human-made' structures did not seem to affect the Roman perception of water.

Recently archaeologists believe they have found the primary spring source of the Aqua Traiana. Beneath a small Christian chapel an ornate chamber was uncovered, decorated richly in Egyptian blue with intricate brickwork; it would have been befitting for an imperial visit. Indeed, it is possible that Trajan may have inaugurated the opening of his aqueduct in person. Regardless of this, the discovery seems to conform to the description of the Virgo's source, in that the function of the aqueduct is mixed with painting and sculpture, creating a celebration of the water source (Taylor, 2012). Undoubtedly the Traiana's source would have been even more impressive in its prime, with statues and other decoration covering the chamber. Other celebrated aqueducts would probably have received comparable attention. The implication, therefore, is that the Anio Valley, where many of Rome's aqueducts were sourced, was littered with these chambers celebrating the birth of water from the ground.

There was also a definite physical relationship between spring and aqueduct that strengthened their association. The standard practice at the selected source was to create many different channels tapping the waters (Hodge, 1992: 77). This meant a series of tunnels being drilled into the aquiferous rock. The water impregnated in the rock would then percolate through the tunnel walls and be channelled into a central basin (Ashby,

1935: 95). The Marcia apparently had numerous such channels that combined to form the main body of aqueduct water. These tunnels reached into the stone and became unseen, enveloped by the natural world. The relationship between spring and aqueduct was thus far more profound than reliance upon a single source.

Furthermore, these aquiferous rocks were not neutral elements in the Roman tradition. The remarkably changeable nature of tufa, for instance (going from extremely soft when first formed to a firmer rock-like texture), is fascinating even with our knowledge today. Its formation by springs (many in the area of Rome) and its absorption of water understandably led to unusual beliefs. In fact, the use of such rock in a building could have been seen as a way of transferring the sacred area of extraction to a new construction. In this vein, Davies and Robb (2002) noted the minor tufa inclusions in Roman temples (and subsequent churches). While the concept of an entirely tufa-based building may seem more noteworthy, the deliberate placement of a few blocks of this rock into a building composed of another dominant material is intriguing. This idea is substantiated, somewhat, by a series of references to tufa by Ovid. In the *Heroides* (15), *Metamorphoses* (8.568) and *Fasti* (2) he makes clear mention of the rock in the abodes of river gods and nymphs. The first of these references even makes a favourable comparison between rough-hewn tufa and the finest Phrygian marble. This running theme of Ovid (Barolsky, 2005) certainly suggests a deep association between rock and water in Roman lore.

All this seems to confirm that, unlike modern water supply systems, the sourcing of water in the Roman world did not divorce it from its natural meaning. Instead, the engineered processes by which humans interacted with water sources like springs, through mechanisms like wells or aqueducts, created greater focus on such values. They also provided the circumstances by which water could be brought deeper into the experience of people living in these landscapes and their own folklore.

Building rivers: hybrid water flow

The passage of water from source to settlement via aqueduct systems has been written about almost entirely in structural and engineering terms. This is no surprise: the great arcades of aqueduct bridges like the Pont du Gard in France are part of this liminal stage of movement through the landscape, and obviously required advanced technical understanding and precise building techniques (Figure 2.2). In these discussions, water is a constant that can be applied to any aqueduct, and in the past the structure dictated its relative success. However, these approaches may as well be talking about water use in the present: the logic is entirely interchangeable, and there is no consideration of the aforementioned local associations with water. If we can talk about the sources of these water management

Figure 2.2 The impressive structure of the Pont du Gard aqueduct bridge outside Nîmes in France.

Source: Image cropped and desaturated by author; original by Benh Lieu Song, https://flic.kr/p/Ht6m8n; licensed under CC BY-SA 2.0, https://creativecommons.org/licenses/by-sa/2.0/.

structures being deeply entangled, hybrid features, then it stands to reason that the movement of that water to the various towns of the Empire could also have more complexity.

Matt Edgeworth (2011) referred to the complex natural and human-made identities of rivers from the medieval through to the modern world. Relationships between communities and their local watercourses led to interactions and changes to protect landholdings, increase productivity, allow new construction projects and facilitate a myriad other local concerns. Such interactions can radically alter waterscapes, and can be witnessed in their most extreme form with an example like the Yellow River in China. One of the two most important watercourses in the country (along with the Yangtze), it is characterised by an unusual amount of sediment, some of which settles along the bed and raises the river higher over time (Ball, 2016: 371). Once the flood season comes, with increased rainfall and snowmelt from higher ground, the river inevitably bursts its banks. For generations farmers have reacted to this by building ever-increasing dykes on the banks, resulting in strange aqueduct-like structures suspending the river at a higher level some 15 metres above the surrounding plains (ibid.: 372). Once these dykes fail, the water escapes and pools into large lakes, unable to return to the original course, leading to a continually changing riverscape.

The Yellow River is the archetype of a river hybrid (Edgeworth, 2011: 77); it is a profoundly altered watercourse that nonetheless still follows natural cycles. While it might be an extreme example, Scarpino (1997: 5) notes that over time many of our rivers have become 'heavily modified, cyborg-like environments'. Due to being largely detached from the extent of changes to our local waterscapes, we tend to polarise our view of them as features solely formed by natural processes or corrupted by human changes. Most people react negatively to this revelation of universal alterations, because of the aforementioned complex disassociation between what is engineered and what is natural; meaning we want safe water in our taps, but this is disassociated from damaging the natural purity of local habitats. The extent to which radical human changes of waterscapes can coexist perfectly well with other enduring values is evidenced by wider anthropological examples. Schumaker (2008), for instance, documents the continuation of water-related myths in Zambia's Copperbelt in the period 1927–1930. A high death rate in the mining community was believed to be due to the actions of a vengeful snake spirit in the Luanshya River. Despite radical altering of the waterscape to eliminate malaria (the supposed cause of these deaths), the legend, with its associated rituals, continued within the local community. Similar sentiments might be suggested for people in the Roman world; certainly there appears to be very little in the way of detachment expressed in the classical sources. People were acutely aware of extensive human interactions with local rivers and water sources, but these interventions did not necessarily erode the meaning-laden aspects of such features.

This blurring of the lines between what could be considered a human-made conduit and what was a natural river can be seen in many different forms in the classical world. The depictions of aqueducts on Roman coins, for instance, follow a very similar format to the aforementioned depictions of rivers (Figure 2.3). Aqueducts like the Aqua Traiana were presented as

Figure 2.3 Drawing of coin issued by Trajan to commemorate the construction of the Aqua Traiana.

Source. Drawing by author.

reclining river deities, the only difference usually being a surrounding structure. This reaffirms that there is a continuation of meaning and associations from the natural source to passage through a human-made conduit. The water of the aqueduct is not suddenly devalued; instead, it remains a river but passes through a different channel. Moreover, the emphasis is undeniably placed on the water, with the surrounding structure appearing to enhance the reclining deity. The reference above to 'cyborg-like' watercourses is fitting, with the addition of the mechanical aqueduct structure allowing the river deity to extend his/her influence far beyond its original limitations.

As with the Yellow River in China, there are numerous references to flooding and changing river flow as an area of legal consternation in the Roman world. In the *Corpus Agrimensorium Romanorum* a passage attributed to Frontinus (*Corpus Agrimensorium Romanorum*: 8.5) notes the example of an old riverbed which, due to the creation of new land deposits through flooding, had its flow channelled in a new direction. Most rivers were public property, but a nearby landowner could now try to claim the old riverbed as part of his land; this would inevitably lead to a dispute with the person who had been disadvantaged by the new river course. Similarly, Urbicus (*Corpus Agrimensorium Romanorum*: 39.28) highlighted the disputes that arose when flooding deposited soil and other materials on nearby land. Seemingly there was often disagreement as to whom this new soil belonged, or who was responsible for removing the waste, especially if it was blocking the river. In the face of such violent fluvial action, many authors emphasised a common practice of making the river and its surrounding area a neutral zone that was public property (*Corpus Agrimensorium Romanorum*: 41.35). This meant that if flooding changed the landscape, it did not necessarily inconvenience private property. In some ways this would have created a type of legal liminality for such watercourses, with no firm sense of associated ownership. On this level, rivers could be labelled as hybrid features, neither public nor private, and requiring sustained human input in response to their natural cycles.

In the same fashion as rivers, the water of aqueducts was legally complex, with a similar mix of private and public elements. Some say the water of aqueducts was not as public as that of an equivalent river. Bruun (2012: 23), for instance, notes that while anyone, within reason, could use the water of a public river, the illegal tapping of aqueduct water was not deemed acceptable. But there may have been a great deal of variability on these issues depending on the local context. For instance, it has been suggested that the water of an aqueduct belonged to the town that constructed it (ibid.), in which case there may have been many legal diversions of the water to supply various influential people or groups. It seems likely that protecting the strength of water flow was one of the leading priorities of such a policy, and in this way it very much reflects the Roman interest in maintaining rivers, noted above.

The practicalities of aqueduct upkeep may be one of the most important overlaps between these conduits and their local waterscapes. The Roman land surveyors and the Theodosian Codex both suggest that during Late Antiquity landowners with aqueducts running through their property were required to pay less tribute, in lieu of being responsible for the maintenance of the conduit on their land. Bruun (ibid.: 22) notes that this may have been a new development in later periods, but it also might just be the earliest evidence we have of long-running practices. There is a fragmentary inscription from Italy with a much earlier date, possibly documenting similar practices, but not a great deal of other direct evidence (ibid.). Of course, it is entirely possible that this was more prevalent in provincial contexts like Britain, where central urban authorities probably overlapped with influential local private landowners and there was less recourse to a defined authority like the 'water commissioner' of Rome.

The *Lex Rivi Hiberiensis* (Lloris, 2006), found in Spain in the early 2000s, tells us of laws relating to a tributary of the river Ebro, which was used as an irrigation channel. The Riuus Hiberiensis was human made but, at the same time, physically linked to the River Ebro, acting like another natural tributary. The law outlines how the surrounding community was required to undertake general maintenance of this canal, including annual cleaning of the conduit. This was part of a series of regular events that paired the local administrative offices with agricultural cycles. The purification of the fields took place at the end of May and was possibly the end of a term for the local *magistri pagi*. The date set to begin cleaning the canal was seemingly 15 July, coinciding with the Ebro's low-water period, which today separates the irrigation period for cereals and that for olives, vines and vegetable crops (ibid.). This then became the start of a new term for local *pagi*.

Consequently, it is viable to postulate that such events may have played a role in a more extensive agricultural celebration, and were seemingly part of an annual calendar of renewal for the community. The entanglement of ritual calendar, practical upkeep and political structure in such action shows how complex the movement of water was in the Roman period. In practical terms, the Riuus Hiberiensis probably bore similarities to many of the aqueducts we find in Britain (discussed in the next chapter). Social dimensions of the cleaning of these conduits have been understated by archaeologists. It is feasible to see such activity moulding group identities along the course of an aqueduct or canal, as people were united by the prime directive of maintaining the structure. Moreover, in specific landscapes it could have replaced or been integrated within pre-existing practices. This would heighten the meaning for an aqueduct as it terminated in a town: not only did it have the original meaning of its source, but it also symbolised the collective endeavour of communities and the close relationships they held with the local landscape surrounding urban centres.

Similar engagement with Roman ritual practice can be detected by analysing the aqueducts of Rome. Frontinus, for instance, informs us of the controversial introduction of the Aqua Marcia. According to him, the Sibylline Books had to be consulted, and it was initially deemed not right for the Marcian waters to be introduced into the capital (Frontinus, *De aquaeductu*: 1.7). If one portrays the aqueducts as purely functional, this episode is slightly puzzling. There seems no reason for these oracular projections to deny the supply of fresh water, so at first glance it seems to be an example of political rhetoric. Yet if one emphasises the cultural associations of water, discussed above, then the first waters brought into Rome from the outside would have had massive significance. In this light it is understandable if a conservative document (which all such codified principles become over time) was resistant to such dramatic change. As discussed above, Rome's waterscape was fundamental to the religious traditions of the city and enshrined within the mythic tales of the ancient kings like Numa Pompilius and Tarquinius. Bringing new waters to this area, laden with their unique meaning, was a profound act. Similarly, when describing the foundation of the Aqua Virgo, Frontinus tells us that the waters of the said aqueduct first entered the city on 9 June (ibid.: 1.9–10); this coincides with the Vestalia, a significant occasion in honour of the goddess Vesta. As the name implies, the goddess was the titular deity of the Vestal Virgins. It has already been mentioned that this group had a ritualistic involvement with water; coupled with the aqueduct being their namesake, surely this must be more than a coincidence. It seems unlikely that the Virgo would have been opened quietly while people celebrated the ongoing festival. Recognition of such a conjunction would suggest some intentional and culturally charged overlap, and hence incorporation of the aqueduct into an event on the ritual calendar.

The vast amount of water that was brought to the city through the aqueducts was eventually channelled into the Tiber through the Cloaca Maxima. It was thus a structure that unified these disparate waters of Rome and merged them with the Tiber. Moreover, the entanglement of 'human-made' and 'natural' water is reflected in the development of the great sewer. Initially it appears likely that the drain followed the course of a small stream running through what would later become the Forum (Hopkins, 2007: 2). As time moved on, this feature was given a structural framework to help drain the flood-prone areas close to the Tiber. Pliny the Elder (*Historia Naturalis*: 36.105) notes how the early Cloaca Maxima collected water from seven tributaries. Its origin as a stream is echoed in the winding course of the feature, which remained unaltered despite later renovation in the Republican and Imperial periods; the total length of the sewer was 1,600 metres, but the actual distance covered was closer to 900 metres (Aldrete, 2007: 171).

Given these attributes, it should not be surprising that the sewer was incorporated in the religious and mythical landscape of the city. In the Forum Romanum, for instance, there was a shrine to Venus Cloacina, a deity whose name is an unmistakable link to the sewer system (Malacrino, 2010: 174). There is also the vivid example of the Bocca della Verità, which, in contrast to its modern associations, was once a monumental sewer cover depicting Oceanus (Hopkins, 2012: 98). Many links have been made between the course of the Cloaca Maxima and other visible religious structures; Holland (1961) and Coarelli (1986) are the most prominent of these, suggesting a distinct link between the sewer and veneration of Janus. The Imperial-era adjustments to the sewer may have been consciously used to relate the feature directly to important structures above ground. A detour close to the Temple of Minerva, for example, was apparently constructed in the same stone, *lapis albanus*, to that found in the temple itself and was cut to similar dimensions (Hopkins, 2012: 190). The rarity of such alterations to the course of the Cloaca makes this selection of material conspicuous. These observations reinforce the view of the urban waterscape as something rich with religious and symbolic meaning, and begin to break down the assumption that such water is distinct from sacrosanct 'natural' water. The sewer appears to be a celebration of the confluence between the water used within the city and the wider water of the Tiber. The meaning-laden nature of the water in the sewer even seems to have survived the very obvious physical impurities and sensory associations that it would inevitably have borne by the time it met the Tiber.

It is also worth noting here that many of the great towns of continental Roman Europe had both grand aqueduct systems and already violent waterscapes. Aldrete (2006) suggests that catastrophic flood events happened in the city of Rome on a relatively regular basis throughout the Imperial period, a time when the monumental building projects in areas close to the Tiber were at their most ambitious and the water supply at its most regular. Cologne (Colonia Claudia Ara Agrippinensium) was built on the banks of the River Rhine and, as a result, was always at a high risk of significant flooding (Grünthal et al., 2006). Even with current prediction and control methods, in 1983, 1993 and 1995 the Rhine broke its banks to wreak havoc upon the city (Disse and Engel, 2001). These are just three years in a continuous history of such events that characterise the geographical location back into prehistory (Herget and Meurs, 2010). The Eifel Aqueduct, which supplied the Roman town, is known as one of the longest conduits to have been built in antiquity and stretches from the distant mountain range of the same name, some 95 kilometres away. This considerable endeavour is remarkable enough alone, but it was supplementing an already impressive network of local spring sources that had been tapped to supply the city with water.

Another example is the city of Lyons (Lugdunum). It was the capital of the three Gaulish provinces as well as being closely tied to Augustus and the imperial cult, so it was a central location of administration and religion, complete with a dedicated sanctuary to the Gaulish provinces (Goodman, 2007: 81). The four aqueducts that supplied the city have therefore been attributed primarily to the wealth of the settlement. Lugdunum was constructed on the important confluence between the Rhone and Saône Rivers, both of which are known for high floods, making the town something of a central point in the waterscape. Certainly the numerous aqueducts seem to add to this feeling of centrality, fanning out in all directions and bringing the waters of the surrounding area to a fusion point. This is mirrored in the prominent Roman towns at Arles and Vienne. The former was characterised by its proximity to a powerful river and consequently beset by frequent violent floods (Bruneton *et al.*, 2001); there is even a suggestion that Roman building around the river made these conditions worse (Allinne, 2007). The latter was located close to a bend in the Rhone, at a large confluence of many rivers. Both had a significant water supply; in the case of Vienne, this has been estimated at potentially 11 aqueducts (Taylor, 1997: 42).

It all suggests a symbolic and ritual productivity to human-made engagements with dynamic waterscapes, with changing pathways of people and water a prime interest within Rome. Purcell (1990) notes how the idea of a 'productive landscape' in Italy often had more to do with symbolic processes of transformation. So, for instance, productive villa landscapes were not always located in places where a maximum agricultural yield could be attained. Sometimes distinctly unhelpful, often waterlogged, locales were selected and then transformed. This would be agricultural production, but it was set against the odds of an unfavourable setting. Thus the ideological potency of the villa could be greatly enhanced. This aspect of Roman thought is explored by Cicero in the discussion of his villa in Aripinum. In *De Legibus* he writes about showing Quintus and Atticus his private villa. Atticus describes the villa as being located on an island in the River Fibrenus, 'whose divided waters leave its verdant sides, and soon re-join their rapid currents'. Furthermore, he adds:

> The river just embraces space enough for a moderate walk, and having discharged this good-natured office, and secured us an arena for disputation, it hastily precipitates itself into the Liris; where, like those who ally themselves to patrician families, it loses its obscure name, and gives the waters of the Liris a greater degree of coolness.
>
> (Cicero, *De Legibus*: 2.3)

The Fibrenus is described as a fast-flowing river that has a noticeable coolness, and when it combines with the Liris it gives the larger river new properties and characteristics. Indeed, both rivers are given individual identities, with the Fibrenus representing an aspiring *novus homo* and the Liris an established patrician. The river beyond the confluence may retain the name of Liris, but the flow is altered and enhanced. Cicero appears to be utilising this charged waterscape to present his ideological image; he was a *novus homo* who had to make careful alliances with established patrician families to cultivate his power. This is an example that shows the symbolic productivity of such waterscapes and how they could be artfully directed to project power. The incorporation of the villa into this confluence zone suggests, again, the blurring of a distinction between the natural and the human within Roman thought. Once more the implementation of human aspects of engineering only seems to strengthen the associations with the water. While reclamation and drainage of water are often cited as the primary exhibition of this symbolic productivity, perhaps the addition of more water to already dynamic waterscapes provided similar opportunities. After all, it is becoming increasingly clear that the selection of settlement sites for Roman towns often appeared to have defied practical rationale, as they were located deep within flood-prone landscapes.

The hybrid baths

All this goes some way to showing the remarkable overlap between 'natural' water and the municipal structures that we have come to see in separate terms. However, if there is one structure we associate most strongly with water in central urban locations in the Roman world it is unquestionably the bathhouse. In many ways these features represent the archetypal Roman building (Laurence *et al.*, 2011: 203); in the northern provinces they are the backbone of any debate that endorses transformation and alignment of a temperate European Iron Age culture with a Roman identity. They resonate so strongly with the Mediterranean, in both purpose and physical form, that it is extremely tough to escape traditional 'Romanising' debates. The logic behind their uptake frequently returns to a compulsion to 'keep up appearances' and embrace an incoming fashion from distant Rome (Potter and Johns, 1992: 103; Rook, 1992: 5). Subsequently, a lack of bathhouses in any area of the Empire has been conceived as indicative of a disengagement with Roman practice (Laurence *et al.*, 2011: 203). As such, we are left with ideas of the great unwashed 'barbarians' clamouring to embrace a defined bathing process because it represented a form of cultural advancement. Too often this decision is portrayed as a given – an approach which unfortunately betrays our lopsided cultural alignment to the 'idealised' Roman period.

Bathhouses throughout the Empire have developed a sort of monolithic identity for archaeologists. As mentioned, they are a phenomenon we see spread out from the Mediterranean; thus writers have been quick to give them a universal cultural value. Part of this cohesion between the vast numbers of baths that have been discovered is the ascribed value of leisure. By its very nature the term is general, and has come to embody a requisite part of life in the modern West. Consequently, the concept is crucial to the way we define 'civilisation' and must have a focal point in the past to satisfy the continual parallels we make between the Roman town and its modern successor. As numerous and often large structures, the bathhouses are a suitable building type with which we can accommodate these sentiments on a grand scale. They remain at the heart of the popular image of Romans relaxing and enjoying their free time and the pleasures their culture provided.

There is a sense in Rome that the baths were conceived as places in which to relax and enjoy life; the increasing move towards providing space for recreational activities in the capital is evidence of this point (Fagan, 1999a: 76). Laurence (2010: 67) highlights how the baths had a unique soundscape of pleasure – masseurs slapping flesh, men grunting during exercise and people splashing in water. At the same time it is acknowledged that there was still a place in society for the more traditional 'stoical' baths linked to famous figures of the past, which individuals like Seneca may have preferred (ibid.: 66). This observation highlights the subjective nature of the terms 'pleasure' and 'leisure', and also the way that the baths accommodated various interpretations of the theme. While it is tempting to identify more with the leisure activities so well attested in imperial *thermae*, there is still a sense that some bathhouses accommodated an entirely different definition of leisure. Subsequently, one wonders to what extent these descriptions befit northern provincial bathing habits, where traditions of leisure are hard to detect in the archaeological record.

In any bathhouse there were a series of specific rooms which were part of an accepted bathing routine. The *caldarium* and *frigidarium* were pivotal rooms found in the majority of public establishments. These stock characteristics of baths are well versed elsewhere (e.g. Nielsen, 1993), and it is not necessary here to document fully the specifics of the accepted bathing process.[4] While these considerations form a part of interpreting how these buildings were used, it is limiting to apply them over such a vast swathe of geographic territory and time. Indeed, with smaller provincial examples people are often more willing to emphasise the conformity to these basic routines (ibid.: 92). While the construction of a particular style can be traced over this large catchment area, the way people experienced each bathhouse would doubtless have been somewhat different. The 'secondary' associations of these structures could be more telling of their purpose (Fagan, 1999a: 10). What is more, these secondary associations actually place water at the heart of varying different processes, further displaying its capacity to become an influential medium in expressing multiple hybrid dimensions. If the source and passage of water to a city can be seen as fundamentally linked to meaning-laden local practices, the bathhouses were monumental confluences of such beliefs.

Healing water

The medicinal aspect of bathing in water has been transmitted in the classical sources, and similar sentiments have flourished in the modern world. In part this has been through the identification with the classical world promoted in recent centuries, but also in the popular idea of the 'natural cure'. The regenerative powers of thermal (or spa) settings are constant with our modern conception of a healthy lifestyle. The utilisation of natural thermal sources in the Mediterranean played a pivotal role in the uptake of the Roman-style bathing tradition. The volcanic area of Campania was a particular heartland for these natural conditions. The Phlegraean Fields, to the north of Naples, are still active areas for heated spa springs (Nielsen, 1993: 21). Within this region, Cumae and Baiae were places where thermal water had a focal role in the creation of settlements. The latter is notable for its numerous bathhouses with medicinal functions (Jackson, 1990: 5). Celsus (*De medicina*: 2.17.1) recommended the sulphurous sweat-baths 'in the myrtle groves above Baiae'; Pliny the Elder (*Historia Naturalis*: 31.2) notes that nowhere were there waters that delivered more 'variety of relief'; even Strabo (*Geographica*: 5.4.5) mentions the hot springs being 'suited ... to the cure of disease'. This utilisation of such places has been mentioned previously, and many settlements throughout the western Empire were named after their thermal spring waters, not least the prime British example of Aquae Sulis (modern Bath). With this in mind, it is likely that some of these bathing facilities were known primarily as places to frequent if one had an ailment or illness.

This tradition of thermal baths is somewhat distinct from the typical structures that we see in almost all Roman towns throughout the western Empire. Yet it is worth acknowledging that the 'artificial' heating found in the vast majority of urban bathhouses probably owes its origin to the imitation of natural thermal water sources, and may have offered similar properties beneficial to health. The literary tradition ascribes the invention of these water-heating systems to Sergius Orata, an oyster grower on Lake Lucrinus in the Phlegraean Fields. He introduced a furnace to heat water before running it under his artificial oyster beds (Pliny the Elder, *Historia Naturalis*: 9.168). Recent analysis has cast doubt on the reliability of this story, despite it forming the consensus origin for Roman bathing into the twentieth century (Fagan, 1996, 2001). The reality may well have been a gradual spread of Greek-style bathing, together with an affinity for such activity stemming from rural practice. Folk remedies, which advised sweating close to a kitchen stove to relieve mild illnesses, are one potential source of the tradition (di Capua, 1940; Fabbricoti, 1976). This is obviously interesting, as it implies medicinal influence in the history of artificially heated baths. In line with the general direction of thinking in this book, the modern idea of this water being 'artificial' did not necessarily preclude it from having powerful local meaning.

Despite not being centred on thermal sites, many baths still enjoyed a reputation for the medicinal qualities of their water. For some physicians the cold baths were just as important as thermal ones in the treatment of various maladies. Even Augustus, suffering from abscesses of the liver, was advised by the physician Antonius Musa to try cold-water therapy (Suetonius, *Vitae Augustus*: 81). The fame of Asclepiades of Bithynia is well observed in the classical sources, and from a general historical perspective on the genesis of modern medicine. It is interesting to note that he was principally associated with the use of cold baths to heal illness (Fagan, 1999a: 98). The term *pensilia balnea* ('hanging baths') is often used in association with his methods, and has been interpreted as a reference to the preference for gravity-fed establishments (ibid.). The implication is that there was significant healing value to a regular civic bathhouse (rather than hydrotherapy being exclusive to thermal sites).

The testimonies of Pliny the Elder (*Historia Naturalis*) and Celsus (*De medicina*) on the matter appear to confirm this assumption. The work of Fagan (2006), documenting the references of these authors to medicinal bathing, captures its huge role in the treatment of illness during antiquity. The first thing that must be emphasised is that both authors are predominantly writing about the everyday bathhouses that we see all over the Empire, rather than the few exceptionally grand structures that so often come to dominate the discussion (ibid.: 192). Celsus alone makes 81 references to bathing in his general account of medicine (ibid.: 205); the majority of which allude to the treatment of relatively severe afflictions rather than more minor muscular issues one might expect. Indeed, taken together the two sources recommend bathing to alleviate symptoms of cholera, rabies, the common cold, complaints of the male organs, diarrhoea, difficult childbirth, epilepsy, eye complaints, fertility problems, fevers, flatulence, gangrene, gout, headaches, itching, jaundice, leprosy, lice, abscesses, palsy, poison, psoriasis, urinary problems, wasting illness and mouth ulcers. Also, there was a more general sentiment that the baths could *prevent* illness and be part of an essential routine for individuals wishing to maintain good health. This is reinforced by epigraphic evidence; an advertisement found at Lugdunum (Lyons), for instance, references the 'healthy little baths' (*CIL*: 13.1926). Many similar examples can be found in the eastern provinces, where the epigraphic record is more considerable (see Fagan, 1999a: 90–91).

It is also worth mentioning that such bathhouses could have been a place where one would find a physician. In many ways this assertion is more conjectural than factual; the actual evidence for doctors at the baths is mostly based on oculist stamps or medical tools found in provincial settings. As Fagan (ibid.: 92) points out, it is an assumption based more on the suitability of the location and the lack of alternative evidence to show this sort of activity was taking place elsewhere. Added to this, the provision of a consistent source of running water would surely have been a requirement for specific medical procedures. The bathhouses were the primary beneficiary of the water supply in most provincial cities, and therefore it makes sense to centralise essential water requirements in these buildings.

Regardless, the sheer breadth of the associations between bathhouses and health is telling. The sources mentioned above give the impression that the act of bathing (in an ordinary civic bathhouse) was seen as something of a cure-all. Even an author like Pliny the Younger, who was not focused on such matters, makes reference to sick men dreaming of baths and springs (Pliny the Younger, *Epistulae*: 2.8.2, 7.26.2), and notes his personal use of the baths to solve an eye problem (ibid.: 7.21.2). Indeed, these beliefs appear to have been widespread across society; the universality of illness would naturally have played a role in the dissemination of such views. Yet the number of different associations with bathing suggests there was a great degree of variability and local difference. This is what one would expect considering the similarly broad healing properties of the springs/water sources from which the bathing water was procured. As a result, there is room to suggest that baths could have had significant associations with healing, but these connections did not necessarily come purely from central doctrines.

Sacred water

These healing properties of bathing were naturally entwined with ritual and religion in antiquity. In this regard there is no shortage of examples to reference if one is covering the aforementioned thermal baths. In Britain, for instance, the bath buildings of Aquae Sulis are overtly religious; the settlement grew up around a *temenos* and thus was always predominantly a religious centre rather than a typical town (Dark, 1993). There are similar examples throughout the western Empire (see Derks, 1998; Yegül, 2010: 51). While there is no need to refute the value of such places in the Roman world (after all, they often exhibit vivid evidence of syncretist water veneration in association with bathing), one must wonder whether they have contributed to us unfairly judging other bathhouses that fall outside these celebrated locales.

The 'everyday' bathhouses found in all the significant Roman towns of the western Empire are often presented as secular structures (e.g. Nielsen, 1993: 146). Nevertheless, within them there was invariably religious imagery, usually in the form of sculpture. The likes of Asclepius and Hygieia were often preferred, due to their role as healing deities (ibid.). Furthermore, these religious figures were not an incidental decoration that could have been ignored; in the Baths of Caracalla gilded colossi of Asclepius and Hygieia loomed above the bathers (Fagan, 1999a: 88). Nor were these associations limited to Rome; excavations at the Hadrianic baths at Leptis Magna have discovered at least six depictions of Asclepius (ibid.: 89). There was also a collection of more ambiguous deities like Neptune, Mercury and Fortuna. Some may say that Neptune is far from obscure, but the universality of water means that he would be relevant to people from all over the Empire; as such, his appearance could be representative of local beliefs despite his Roman presentation. Similarly, Mercury and Fortuna have an undefined value which means they were worshipped for many different reasons. The latter is sometimes referred as to as Balnearis, the patron goddess of baths (*CIL*: 2.2701). This particular association probably derives from the medicinal uses of bathing, mentioned above. In particular, Fortuna seems to fit with the idea of preventive measures in avoiding illness.

It is important to add that the respect paid to these deities was not just expressed by depictions within the bathhouses themselves; sometimes these structures were monumental oaths to healing gods made on the part of wealthy benefactors. The restoration of the Hadrianic baths at Leptis Magna was such a statement from Cornelius Attax Marcianus and Lucius Appius Amicus Rufinianus (*AE*: 1925.105). So while the function could feasibly be interpreted as entirely secular, the motive is grounded in religious belief. Subsequently, the construction of the bathhouse was perhaps almost akin to a votive deposit on a massive scale.

With this observation in mind, outside the thermal bathing centres many religious sculptures in bathhouses have been interpreted as superficial decoration (e.g. Nielsen, 1993: 146). This is not necessarily surprising because, in contrast to somewhere like Bath, the common bathhouse may have held sculpture but possessed little in the way of overt religious material culture. Without curse tablets or closely associated temple precincts, it is hard to make an archaeological judgement against the mostly secular definition of these buildings. In this regard, it is important to acknowledge that an individual is probably more personally involved in the bathing process than would have been the case with many religious rites. The act of immersion inherent within bathing is more involved than any of the water worship mentioned in this book: it represented a sustained corporeal experience rather than the deposit of an item in a particular place. Indeed, the temporary deposit of one's body could have been a significant ritualistic statement in itself, in which case it would be undetectable in the archaeological record but entirely profound for the person involved at the time.

The changing nature of the water, in parallel with the movement of an individual, through the baths holds significance. It created a sense of liminality, similar to other watery circumstances that are highlighted throughout this book as places of veneration. This may be an element of the bathing process that is affected by our overfamiliar approach to the experience. Frequenting the baths may have been a regular activity for many people, but it was also an undeniably transformative experience. On every level, these structures dealt with a changing of states. The various rooms represented different elemental expressions of water: the *caldarium* a heated thermal vault, the *frigidarium* a frigid and arresting shock. Steam was created in the hot rooms of the baths, along with condensation on the walls. In the colder areas, perhaps in winter one would see the vapour from hot bodies rising to the ceiling; occasionally the icy conditions could have filtered into the internal pools themselves. Indeed, the baths provided a microcosm of the many states of water in nature, with the result being drained away into a nearby river. If the aqueducts supplying these buildings can be seen as rivers, the baths themselves displayed water in its other elemental changeable forms. There is ethereal majesty to the process that cannot be undervalued because it fundamentally links to myth and tradition. Yegül (2010: 80) refers to the frequent occurrence of references to Vulcan and Neptune in the ancient literature as poetic metaphors to describe fire and water. Their elemental contrast is inherent within the processes of the bath, and thus places the practicalities of maintenance into a mythic framework.

The senses of the individual would have been challenged continuously with the changing conditions. The steam of the hot rooms could have obscured and manipulated identity, possibly allowing activities of a clandestine nature to take place. The sounds of a busy bathhouse were similarly deceptive. Issues of self-identity were also potentially dealt with: the reflection of one's image in the waters could be a reminder of perceived health (or ill health), and the disrobing process may unveil issues of self-image. Laurence (2010: 69) refers to the importance of physiognomistic observations in the determination of character in the Roman world. The power of an individual could be projected by careful consideration of cultural affectation in the bathing process, and thus may have had a marked effect on the perception of that person within a community. This must also engender a strong sense of local preference and style, again raising the question of provincial settings.

Transformation is something that seems to have been influential in Roman thought (Ferris, 2007: 121). Famous works such as Ovid's *Metamorphoses* are clear examples of the role it played in a more extensive mythic belief system.[5] Indeed, deities within the classical tradition are constantly changing form to suit their own needs. As we have seen, the rivers of Italy were also shown to be changeable, sometimes flowing through the land and at other times taking the form of figures that could interact with humans. The possible importance afforded to calcareous tufa, discussed in the previous chapter, is another example of this emphasis on transformation. In this regard it is worth dwelling on how different sources of water reacted to the various processes of the bathhouses. Heat can on occasion bring out specific properties in water, which in turn could have affected the perception of the bathhouse.

The extent to which municipal bathhouses bore religious meaning might be borne out in the propensity for these structures to be developed into churches in Late Antiquity. Submersion within water is of course a notable part of Christian belief, and is displayed in the tradition of baptism. Moreover, there are some traces of this Christian rite being influenced at some point by the widespread bathing traditions of the Roman Empire. The octagonal baptistery form, for instance, seems to have been adopted from the *frigidarium* of the baths (Nielsen, 1993: 145). Certainly, during Late Antiquity the bathing tradition was endorsed by the Church fathers, even if they disapproved of some aspects of the process (ibid.: 111). A great many bathhouses were later transformed into churches, incorporating the water supply into the religious processes of the new building. Monastic water supply is a rather intriguing area of study. These constructions often utilised earlier urban water supply systems, the same ones that would have fed the Roman baths, but they are more readily conferred a clear religious meaning. It seems likely that they appropriated past associations as much as creating a new mythos surrounding water.

Communal and political water

Considering the varied associations with water established in this chapter, it should be unsurprising that people wanted definitive places to interact with it. The bathhouse gave people of all different backgrounds a chance to engage with water: the large *thermae*, particularly the massive examples in Rome, were frequented by a broad range of society and were inevitably a point of contact in an increasingly populous urbanised world. As Fagan (1999b: 33) points out, there is nothing explicit to counter the argument that even slaves were allowed to frequent these structures as clientele. Indeed, there is no direct evidence implying that these lowest members of society were banned from any establishment, or were required to bathe in special places. There is a line of thought suggesting that there could have been a sort of unwritten conformity to the particular times people visited the baths. Laurence (1994b: 108), for instance, mentions the possibility that only the elite would have had free time to use the baths at the optimum times. However, it is hard to apply such logic in a general sense, especially outside Italy; thus it seems reasonable to suggest that the social diversification within a bathhouse at any one time could have been marked. This facilitated interaction between people from very different backgrounds for many contrasting purposes. Clearly this is why bath-houses could often develop a link to shady social interactions, such as prostitution, scheming or petty theft. The presence of the latter is remark-ably well preserved in the archaeological record, with numerous curse tablets relating to the theft of clothing or other personal accessories (Fagan, 1999a: 37). There was also undoubtedly the potential for more nefarious dealings between local elites that could have represented organised crime.

As a point of confluence for many different types of people, it was inevit-able that the bathhouse would become an arena for economic activity. There is widespread evidence of shops incorporated into these buildings, and the sale of food is of particular note.[6] Indeed, Martial (*Epigrammata*: 12.19.1–2) portrays one Aemilius as eating eggs, lettuce and fish in the *thermae*. A *graffito* from outside the forum baths at Herculaneum confirms this activity; it is a price list that documents drinks, hog's fat, bread, cutlets and sausage (*CIL*: 4.10674). This sort of evidence makes one imagine the baths are a central point for street-food vendors, the like of which are still seen in many Italian towns and cities today. Incidentally, it is worth noting that this type of food invariably maintains strong links to local com-munities: street food consistently represents local tradition and practice even in modern times. This, importantly, adds a layer of diversity to the bathing experience beyond its representation as a unified phenomenon.

In a less direct manner, the more organised activity of 'dining' was also part of the bathing experience. Fagan (1999a: 22) notes how Martial makes numerous references to meeting dinner guests at the baths, as a type of prelude to the actual event that presumably took place in his private residence (e.g. Martial, *Epigrammata*: 11.52.1–4). Sometimes this is a matter of spontaneity, where the idea of dining is a product of meeting a certain group of people at the baths. Laurence (1994b: 110) mentions how the best time to frequent the baths (due to ideal temperature) would have fitted well into this role as a prelude to dining. Thus by either spontaneity or arrangement the bathhouse formed an essential part of the Roman dining experience. Both authors link the act of bathing to food consump-tion. On the surface this may not seem like a particularly profound con-nection; yet when one applies this logic to the provinces it becomes a more notable phenomenon. In temperate Europe there was a distinct feasting culture (Murray, 1995; Arnold, 1999; Poux, 2000; Ralph, 2007) that was highly ritualised and involved in the formation of identity. As mentioned in Chapter 1, votive deposits of food in water could have been an increas-ing feature of the Late Iron Age. The potential integration of this type of activity into the function of a bathhouse immediately alters the parameters within which we may assess the experience.

The social diversity of the clientele meant the construction of new baths gave a political boost to any benefactor. The *thermae* of Rome are monumental examples of this: Agrippa (for Augustus), Nero, Titus, Trajan and Caracalla are among the more prominent figures who constructed these massive urban features (Figure 2.4) as a means of influencing public opinion. Nero even tried to use his baths on the Campus Martius to engineer fundamental social change; he was the first to expand the activities that took place in these buildings with the introduction of rooms for lectures, poetry reading and music (Nielsen, 1993: 46). He was probably trying to broaden the role of the baths to encourage more of a Greek gymnasium style that encompassed many different activities (Tamm, 1970). Trajan took this ideological element even further, with the construction of what DeLaine (1999a: 70) calls 'an entirely self-contained cultural world'. Such a structure undoubtedly realigned the social traffic of Rome away from traditional centres such as the Forum Romanum. This imperial patronage of bath buildings was also prominent in the wider framework of Italian towns. Records show the construction or restoration of bathhouses by Augustus and Gaius at Bologna (*CIL*: 11.72), Antoninus Pius at Ostia (*CIL*: 14.98) and Tarquinia (*CIL*: 11.3363) and Caracalla at Pinna (*AE*: 1968.157). The actions of the emperor naturally had a filter-down effect, with other figures of importance coming to see the patronage of baths as a genuine source of prestige within their communities. As DeLaine (1999a: 72) notes, important senators, such as Pliny the Younger at Como (*CIL*: 5.5262), were more than willing to follow this example. Moreover, there is even a record of Ceionia Demetrias, Commodus' concubine, restoring baths at Anagni at her own expense and subsequently receiving a statue

Figure 2.4 The baths of Caracalla.
Source: Picture taken by author.

in Anagni's forum (*CIL*: 10.5918). The whole process became a complex system of euergetism in which both benefactors and the recipient community made substantial gains (Zajac, 1999: 100).

It is important to note the large imperial benefactions because they inevitably led to smaller-scale imitation on the local level. An ambition to ascend up the political ladder may have been part of this; the role of the *aediles* was at one time to oversee the running of the baths (Nielsen, 1993: 125). However, this system could also have given a very local flavour to some bathhouses in Italy and beyond. The vast majority of bathhouses outside Rome (in Italy and the provinces) were constructed by locals, either acting for civic authorities or as private benefactors (Fagan, 1999a: 142). It is also possible that there was no distinction between civic authorities and private benefactors, and that in reality responsibility for construction and restoration fell to the same people. Regardless of this, there is a strong pattern in the epigraphic record suggesting local financing of these structures – something that is inevitable considering the numbers of bathhouses established outside Rome. Many of the inscriptions relating to definite private benefaction refer to unknown locals,[7] but there are also examples of very successful individuals (who had moved away from their homes on imperial business) arranging for these structures to be built in their place of origin. Fagan (ibid.: 153), in his epigraphic sample of the phenomenon, lists a *praefectus Aegypti* building a bath for his native Volsinii (*CIL*: 11.7285), a *primiplaris* and *tribunis* doing likewise at Fanum Fortunae (*CIL*: 11.6225) and even Pliny the Younger leaving money in his will for a bathhouse at his native Comum (*CIL*: 5.5262). These examples show that even for individuals who had travelled away from their native lands, the bathhouse was seen as the appropriate vehicle to illustrate their success on a local level.

These ideas of local influence are particularly relevant when we consider the differing cultures of the provinces. Moving outside Italy but staying in a Mediterranean setting, the archaeological preservation of baths in North Africa is particularly good, and archaeologists have been able to pinpoint examples of clear local variation rather than merely projecting generalised forms. The Hunting Baths of Leptis Magna (Libya), for instance, were constructed throughout in rubble concrete, with vaults and domes providing an architectural profile that was in marked contrast to the more traditional architecture of the surrounding Roman city (Wilkes, 1999: 21). It is important to note, with regard to the culture of benefaction discussed above, that the original plan of this bathhouse was changed, with several new structures being added. These additions were timber roofed and built in orthostat and rubble concrete combinations (*opus africanum*), a local technique (ibid.: 23). Next the internal decoration was given an overhaul, with hunting scenes proliferating. A painted frieze some 1.60 metres high and depicting the hunting of lions and leopards was added in the *frigidarium*. The suggestion is that these innovations could be attributed to a wealthy local individual who possibly procured wild animals for the nearby theatre (ibid.). Neither is this particular bathhouse alone in its local variations.

The so-called 'Unfinished Baths' was named due its series of structural anomalies, such as the lack of vaulting over the cold sector, the absence of a connecting passage between the *tepidarium* and *frigidarium* and the contradictory placement of said *frigidarium* on the western side of the building where it would catch the afternoon sun (Maréchal, 2013: 207). Yet these differences may be due to a different function for this building, despite its shared characteristics with bathing structures found elsewhere (ibid.). The Eastern Baths at Leptis apparently had a kiln installed while still functioning as a traditional bathhouse (Stirling *et al.*, 2001). The key idea to acknowledge here is that an ambitious local personage probably financed the building of these bathhouses, local skills were used in construction and maintenance, local culture is represented through the decoration and even important local manufacture may have been located inside the buildings; hence a Roman concept – that of the bathhouse – becomes a focal point of cultural hybridity within the society it serves.

This is even more pronounced in other instances from North Africa. For example, the Small Baths at Thenae in Africa Proconsularis (modern Tunisia) had a unique layout: instead of a loose idea of sequential rooms, there is a cluster of geometric-shaped rooms packed around a circular *frigidarium* (Yegül, 2010: 150). This means the bathing experience would have been radically different from the usual routine. The circular navigation of the rooms, with their changing shape and the steam from the baths, would surely have had a bewildering effect. This was not about an emulation of a Roman standard, but rather a unique journey carefully curated by a talented designer. While examples of baths from Roman Britain do not appear to be as unorthodox, we do not have the same body of evidence to interpret structural diversity. Instead of automatically rounding our interpretations up to traditional forms and functions, we should acknowledge that in other provinces local variation was a vital part of bathhouses; thus it may have manifested in some form in a British setting.

Hybrid urban water networks

This chapter outlines how the Roman view of urban water networks cannot easily be quantified in terms of just practical consumption or conspicuous waste. This binary perspective, abstracting the interpretation of water supply from the broader waterscape of Roman towns and cities, is a product of our bias rather than an accurate reflection of beliefs held in the past. This acknowledgement is a primary aspect of the hybridity written about in this book. Our modern encounters with water in urban centres emphasise an apparent duality of meaning, with the supply to our homes distinctly separate from various abstract notions of nature (whether that be pure spring water or disease-ridden rivers). It is a state of mind primarily caused by being divorced from the means of production for our water, as mentioned in the previous chapter. Unsurprisingly, this is not the case in the Roman period, when people had a deep relationship with their local natural landscapes born out of necessity and layers of meaningful interactions over generations. The addition of human-made elements to these waterscapes does not appear to have eroded this significance – if anything, more sophisticated human interactions allowed the meaning of the water to be enhanced and performed in new urban spaces. By acknowledging a more hybrid view of the municipal water, we can start to see its importance in the creation and fortification of distinct local identities. While there is no doubt that Roman supply networks were involved in providing water for personal consumption, and in many places in the Mediterranean also exhibit elements of gratuitous waste, these are just two aspects of their function. Yet by placing unbalanced emphasis on these we direct the discussion towards uniform interpretations of worth with very narrow criteria. The evidence from the classical sources, surveyed in this chapter, suggests a more nuanced sense of importance that was ultimately considerably affected by local practices.

It is abundantly clear that wells, aqueducts, sewers and bathhouses had strong religious and symbolic value in the Mediterranean, allowing associations with springs and rivers to be focused in the centre of urban locations. Ritualistic and religious engagement with rivers and springs had never been based on uniform values, but rather a highly diverse accumulation of local folklore. The urban water network of Rome was an integrated part of this folklore, with different mythic backstories and formalised ritual events in line with the various stories associated with 'natural' elements of the waterscape, such as springs. The act of bringing new deities to Rome is a prominent part of wider studies of Roman religion. Historians rightly highlight how the import of Magna Mater, for instance, was a profound action for the people of Rome and changed the religious landscape of the city (Orlin, 2002: 110). Bringing water into the city should be seen as a development in the same vein – it certainly seems to have been celebrated and honoured as such within the classical sources. The primacy of the Roman bathhouse at the heart of urban life could be considered a testament to this sentiment. The diverse local significance of the water led to these structures being malleable sources of local political power, religious reverence, healing efficacy and social togetherness. Unlike the silent water of modern cities, the urban space most synonymous with water in the Roman city was transformed, depending on the meaning-laden water it housed and displayed.

Rogers (2013) and Purcell (1990, 1996) both singled out the political and symbolic currency that water reclamation projects had in Italy and the provinces. Transforming a problematic watery landscape into habitable land was a convincing method by which Roman individuals could express their local power – whether that be on the scale of a town or of a villa. As with all expressions of political power, this could be contentious. The variable legal definitions of water discussed earlier contribute to the complexity, and show the struggle between local authorities and wider imperial ambitions. Water management in the Roman town should be seen as similarly open to various interpretations. By accepting a hybrid aspect to water, we move beyond simplistic reactions to the uptake of characteristically Roman water features. Their creation was not merely about aspiration towards a Roman ideal, or active replication of it. These were structures that could manipulate the perception of local space and thereby create divergent reactions, both positive and negative, depending on the individual experiencing them. It is also possible to see water networks being built to serve very provincial needs, perhaps distinctly at odds with Mediterranean practice.

Embracing this notion of urban water being integrated into the wider dynamic of its local context fundamentally changes the ways we evaluate the reasoning behind their construction and quantify their success. It is always tempting with these structures to suggest that the greater the scale, the greater the symbolic meaning. It cannot be denied that a structure like the Pont du Gard in France or the Aqua Claudia in Rome was impressive. But highlighting the meaning-laden nature of water, and the hybrid effect this had on its associated features, means that scale does not necessarily equate to success. This is an especially important factor when we come to assess the use of water in a province like Britain.

Notes

1 This practice of viewing rivers as personified individual deities is something that can still be seen in published accounts in the early 1600s, examples of which are documented in Hingley's work on Hadrian's Wall (Hingley, 2012: 77–80). The impact of the Enlightenment (see Chapter 1) was likely key in the gradual marginalisation of this practice.
2 There are numerous examples of this, such as Holywell (Huntingdon, Cambridgeshire), which is dedicated to St John the Baptist, stands above a well and existed before 1007; St Bride's, an eleventh-century church just outside London, which was aligned on an artificial well or cistern and cut through the remains of a Roman building; and the 'Old Church' at Cokethorpe (Oxfordshire), which dates to AD 958 and possesses a curative well (Blair, 2007: 377–378). All these are churches of the early Middle Ages, and seem to have potential for interpretation of their importance before the establishment of the Christian buildings.
3 Indeed, there seems a degree of inconsistency (to be expected) in the treatment of wells in the Middle Ages. Some were assimilated into the rationalisation of the saints or the Virgin Mary, while others were frowned upon by Christian authorities (ibid.: 477 and 481).

4 The most basic bathing routine was to visit the medium-heated room, process to the hottest room, return to the medium heat and then onward to the cold room (Fagan, 1999a: 10).
5 In *Metamorphoses* there are numerous examples of transformations involving water. Sometimes this happens when an individual inadvertently offends a river deity – such as when Acteon is changed into a stag by the water of Diana (Ovid, *Metamorphoses*: 3.288).
6 The fact that many bathhouses were close to the local forum and *macellum* would have increased such activity.
7 This group also contains some notable female examples; one Vocina Avita of Taglis (Tíjola, Spain) constructed baths at her own expense, and gave the town a considerable amount of money for the upkeep and perpetual use of the structure (*AE*: 1979.352).

Chapter 3

Water in Roman Britain

Establishing a context for water

Establishing that any incoming 'Roman' traditions related to water were both inherently variable and also deeply concerned with local traditions should force us to re-evaluate our interpretations of the evidence of Roman Britain. There was a definite precedent for meaningful interactions with water in the prehistoric past of many places that would later become prominent Romano-British cities and towns. Moreover, it must be made clear that these sustained pre-Roman engagements with water were not so different from the later developments (or the wider traditions outlined in previous chapters), and should not be dismissed in Romanist scholarly contributions. If we pursue the outcomes of the preceding chapters, Roman constructions, even in their original Mediterranean form, should be considered meaning-laden developments in the relationship between natural context and human urban growth. Moreover, they possessed a high degree of local diversity. Thus the immediate history of different locations in Britain must be considered, because even a radical change away from pre-existing practices could be potentially revealing.

As briefly discussed in Chapter 1, the conceptual divide between prehistoric and Roman-period interaction with water in Britain is quite striking. The only area where there is a degree of crossover is the treatment of wells and underground water. Ross (1968) and Wait (1985) provide the two large-scale surveys of Britain; the former emphasises examples from different periods to support the argument. Both these contributions, plus the significant work of J. D. Hill (1995), emphasised votive items deposited in these features during later prehistory. The same types of interpretations have been made at Roman-period settlements like Silchester and London, and are discussed later in this chapter. Of course, wells were given ritual prominence in later periods (Webster, 1995: 452) and, due to their limited presence in modern urban centres, are more insulated from twentieth and twenty-first century rationalisation. That said, even with this type of feature there can be an overreliance on analysis of technological sophistication to justify importance – something which almost always puts primacy on dialogues of incoming Roman expertise.

Romanists are often dismissive of wells that did not function properly or do not house noticeable votive deposits; a way of thinking that has been vigorously challenged by scholars studying the Iron Age, who have emphasised the importance of deposits that may be described as 'rubbish' (Hill, 1995). There is also the issue of more ephemeral meaning for these features that are easily overlooked by archaeologists but potentially had a transformative effect on the perception of individuals who interacted with them. For instance, there is a strong British folklore tradition of prophetic drumming wells (Cordner, 1946); while it is dangerous to assume a direct link in the Roman past, the drumming wells have a similar sense of independent agency to the self-purifying wells discussed by Seneca (*Quaestiones naturales*: 3.26.6). The act of dressing wells is also very rarely mentioned in the interpretations of archaeological excavations that discover urban examples of wells, but it could have transformed an architecturally humble feature into a central point of symbolic and religious meaning. Moreover, the myriad of potential associations inherent in the actual water is too often just dismissed.

The more monumental the water feature in the Roman period, the more likely it was completely divorced from any traditions or practices that were present in the same landscape in prehistory. As a result, these structures are judged in purely 'Roman' terms centred on issues of scale, technological sophistication and alignment to a universal Empire-wide ideal. It is an evaluative framework that has led to British aqueducts, for instance, being almost wholly ignored within large-scale studies of such features across the western provinces. This exclusion is because many of the examples we know of have taken the form of smaller-scale leats (often from river sources), without the ornate arches and bridges that we see comparatively often in provinces like Gaul. Three primary reasons have been offered to account for this discrepancy. The first is a disparity in the amount of money available to construct large works (Stephens, 1985a: 204). Second, the geography of Britain was a vastly different prospect for Roman builders than that of Gaul, where the dramatic landscapes that characterise many regions demanded dramatic solutions in terms of aqueduct building; the more sedate rolling countryside of Britain required less sophisticated techniques. Third, it has been mooted that there was just not the expertise in the province to execute such projects.

Similar problems continue to define our interpretative framework for urban bathhouses in Britain, which remain reliant on us characterising them as a broadly homogeneous cultural phenomenon. Functions of bathhouses in London have been deemed substantially analogous to those found in Paris and in Rome itself. As a consequence of this reliance on uniformity, we have been mostly content with promoting the idea that people in Britain were building and frequenting baths to 'keep up appearances' with the rest of the Roman world (e.g. Potter and Johns, 1992: 103; Rook, 1992: 5). In any foray into the literature on bathhouses one is sure to encounter recognisable nomenclature lifted from the classical sources: terms like *thermae* and *balnea* are regularly deployed to signpost the importance of a bathhouse. Modern scholarship has often emphasised elements of scale or grandeur as the primary attribute of the *thermae* (Nielsen, 1993: 3; Yegül, 2010: 101), the best examples being the vast structures constructed by imperial benefactors in Rome. The *balnea*, by contrast, are smaller, more rudimentary structures. This definition may be acceptable in the capital, but it inevitably creates problems when analysing the provinces. The baths at Wroxeter and Silchester have both been interpreted as *thermae*, but their direct comparison with the contemporaneous structures of Rome in terms of size and decoration somewhat stretches the definition. From an archaeological perspective, the remains in Britain do not necessarily lend themselves to interpreting this grandeur; internal decoration can usually only be partially assessed and is often assumed.

One can find similar issues in the reinforcement of room definitions such as *caldarium*, *frigidarium* and *tepidarium*. It is standard practice to label bathhouses with these terms, and in the process create a complete and satisfying interpretation of function and experience; at the same time, it can force us into dangerous assumptions based on what we presume should be present. The recent reassessment of traditional notions regarding the spatial distribution of activities within Roman domestic contexts, even in core Mediterranean locations like Pompeii, shows the need for caution (see Allison, 2007). A valid example of this is the excavations in East Stockwell Street in Colchester during the early 1990s, which uncovered a building with substantial foundations (Crummy, 1991; Benfield and Garrod, 1992). The structure had an unusual layout, with a number of surprisingly small rooms and one large room that measured 10.5 metres wide by 27.5 metres long (Crummy, 1991: 9). A central drainage feature within this room, coupled with the considerable foundations, suggested the possibility that this was a bathhouse. Hence the usual terminology was applied, meaning the large space was labelled as a *frigidarium*. This may well have been an accurate interpretation, but there was little interrogation into why this particular bathhouse had such unusual internal dimensions. The implementation of the accepted interpretative framework often takes precedence over any scrutiny of local variability and practice. So pervasive is this way of thinking that even progressive writers (e.g. Mattingly, 2007: 283) who endorse ideas of discrepant experiences highlight that known differences in certain bathhouses are just exceptions to the general rule of uniformity.

Invariably, generalising these structures leads to them being presented as simplistic archaeological plans, with familiar labels for the various rooms (e.g. Nielsen, 1993). As a result we sanitise bathhouses, completely removing any relationship between them and their unique local context. One can look at such a plan and interpret the accepted function and experience of the building without any knowledge of where it was located. It is a process of reductionist interpretation reminiscent of processualist approaches to Roman urban studies (e.g. Hodder and Hassall, 1971). Moreover, it is a way of thinking that lends itself to ignoring the import- ance of diversity and presenting a complete and easily recognisable cultural phenomenon.

As a result of these approaches, Romano-British urban water manage- ment has been defined mainly by failed aspiration to a pan-Roman ideal. However, considering Roman water infrastructure as a more integrated part of local landscapes, as discussed in Chapter 2, it is possible to see these developments as holding a great deal more complexity. Ultimately, similar types of interventions were already happening in the Late Iron Age. Monumental causeway structures, mentioned in Chapter 1, undoubtedly altered the flow and the perception of a river. Even a class of monument such as the pit alignment, which consisted of a series of pits dug to form rows, could have a physical and conceptual effect on the flow of water. These pits have been enigmatic for the archaeologist, lacking any apparent practical reason for their institution.

However, a number of authors have linked these features and water (see Pollard, 1996; Willis, 2006; Rylatt and Bevan, 2007). Rylatt and Bevan (ibid.) note how, in their experience on site, these pit alignments naturally filled with water from rainfall and remained for some time as pools of water. The result is the creation of a reflective monument that separates itself from the landscape by manipulating the water (ibid.: 221). Through examination of pit alignments at Kilvington (Nottinghamshire) and Gar- dom's Edge (Derbyshire), Rylatt and Bevan (ibid.) postulate that the leading role of these monuments was to hold water, thereby altering tradi- tional boundaries within the area. The close association of these structures with rivers strengthens this idea (Pollard, 1996); it suggests their construc- tion in the Iron Age could have been a profound reaction to changing land- scape conditions and a method of defining liminal zones. Communities may have created these monuments so they could be inhabited by a nearby water deity, a traditional force of power, to legitimise land arrangements. This may have affected the perception of other local communities. They could also be an expression of appeasement, to placate a river deity who is encroaching upon the land through flooding. These monuments were a vivid demonstration of the influence a river could exert on its surrounding landscape. They also made it easier to interact with the water, instead of contending with the dangerous full flow of a river.

There might even be a case for arguing that pit alignments 'encased' parts of rivers, albeit in a less physical way than an aqueduct. It has been proposed that some of these pit sequences represent cumulative layers, or additions, around a river (ibid.). Moreover, they are a relatively wide-spread feature of the Iron Age that coincided with areas that later became locations for aqueducts, such as Lincoln (Boutwood, 1998; Willis, 2006). Such demarcation of rivers by pit alignments created a liminal space that protected the physical and symbolic integrity of watercourses, somewhat akin to how a strip of land had to be left open on either side of Roman aqueducts (Frontinus, *De aquaeductu*: 1.127).

So there is a sense that incoming traditions associated with aqueducts in the Mediterranean would not necessarily have been viewed as utterly alien to the inhabitants of Britain; thus the introduction of these structures is not taking place in a cultural vacuum. While pit alignments are unlikely to be found in a Roman town, they are another aspect of a reverence for ground-water that has received little analysis despite being widespread in later pre-history. Perhaps the most intriguing element of this is the sensory appreciation of these pits and the possibility that they could be linked to a nearby water deity. This type of analysis moves us on from the somewhat isolated treatment of votive deposits into the more ethereal spatial links such features could have had in communities of the past. Rylatt and Bevan (2007: 223) quote Bradley (1993: 116), who notes how 'new developments are more secure where they are invested with the authority of the past'. For them, this is the process of how pit alignments were manipulating an older traditional power (in the utilisation of the water), thereby legitimising the new monument and possible land division. This may have supplemented similar arrangements with naturally occurring pools and water channels in the area. However, moving into the Roman period it is highly possible that such processes continued to develop and found themselves expressed in the new urban form. This may have played a part in the creation of a new hybrid sense of place in the Roman town.

Settlements of the Late Iron Age are also known for monumental dyke systems, the functions of which remain unclear. The traditional consensus has been that they were defensive works; yet for the most part these features did not completely encircle settlements and therefore are of debatable value in a defensive capacity. Rogers (2008: 80) mentions that some of them could have been used to direct and channel livestock; the importance of cattle as symbols of power and status may have been a reason for such a monumental undertaking. This raises crucial questions about how people of the Iron Age conceived the connectivity of places within the landscape. Part of the interpretation of dyke systems as outlets for moving cattle is based on the linkage these structures provided between wetland areas. The movement of animals to and from different watering holes would be a practical necessity in a more fluid agricultural setting, where landholding will doubtless have been a sensitive and well-managed matter. Indeed, writers have utilised the term 'hydraulic communities' to describe the pattern of activities in such landscapes (Evans, 1997; Chadwick, 2007).

The possibility that such actions were ritualised and monumentalised constitutes a key development in this area.

All this is very similar to the meaning-laden aqueducts of the Roman world discussed earlier. We can see structures that physically and symbolically link places in the landscape. More than this, they become hybridised features that heighten the flow of things between places; in this case, we have animals. Willis (1999) noted how some Iron Age enclosure systems in the northeast of England seem to have been constructed in direct relation to the surrounding environment. At times of wood surplus, large palisades were constructed as using this material; but when it became scarce in the area, ditches and banks became more predominant, replacing palisades and echoing a 'cleared' landscape of fields. The result is a human-made feature in harmony with its natural surroundings. The development was not a radical new statement, but a change that better expressed contemporary associations with the immediate landscape.

As noted already, the bathhouse may be the urban water feature that is most detached from any previous activity, at least in the modern reader's mind. In many ways we have followed the logic of some of the classical writers and envisaged bathing as an alien practice to the population of pre-Roman Britain. Galen, for example, sums up his view of bathing on the northern periphery of the Empire by stating that the 'ponds and rivers are to them as the bath is to us' (Galen, *De sanitate tuenda*: 1.9). His intention was likely to be pointing out the cultural superiority of Roman practice to the 'barbarians' of the north. Modern writers have tended to reinforce the idea of the baths as a marker of civilised behaviour, diverging from previous activity. Yet putting issues of cultural superiority aside, it is notable that Galen is not stating there was no bathing in temperate Europe; he is merely noting that such activity took place in ponds and rivers. This should be immediately interesting to archaeologists, because these are presumably the same ponds and rivers that are widely attested to be locations associated with important ritual and symbolic practices in prehistory throughout the region. Any bathing practice taking place in the vein Galen describes thus appears to be closely linked to such activities. Of course, it is also a connection that is present in the Roman period. The water of bathhouses did not come from the ether, but was sourced from the same rivers and ponds to which Galen refers. The link between the baths and their immediate context is therefore unquestionably something that needs accounting for when we analyse northern provincial locales.

The reality is that even bathhouses were not constructed in a cultural vacuum in Roman Britain. In fact, there may have been similar features already present in northern Europe. Ambiguous Bronze Age 'burnt mound' features, for example, have been interpreted as a simple form of sauna and equated to Native American sweat lodges (Barfield and Hodder, 1987: 374; Laurie, 2003: 244–245). In anthropological studies these small sweat lodges have been shown to be places where various rituals were performed (Bruchac, 1993). The isolated context of the 'burnt mounds', away from large settlements, perhaps hints at similar rites taking place. While there is less evidence for Iron Age parallels (albeit there are a few instances), it is possible that potential features have been misinterpreted or poorly documented by archaeologists concentrating on the sometimes more vivid evidence of later periods. We should not discount the possibility that such activity was still taking place in the Iron Age; the 'Celtic' cultures of northern Iberia, for example, are known to have used large decorated stone slabs termed *pedras formosas* in heat production. These were part of sauna structures that were likely developed independently of southern Mediterranean traditions (Parcero Oubiña and Cobas Fernández, 2004; García Quintela and Santos-Estévez, 2015). While it might be a stretch to suggest there was a defined pre-Roman bathing culture in Britain, there is indeed evidence to suggest an increasing engagement with water that needs to be accounted for when we interpret the construction of bathhouses in the Roman period.

Ultimately, this chapter is about anchoring Roman-period developments in a long sequence of local water interactions. Bearing in mind the content of the previous chapter, it seems highly unlikely that incoming continental 'Roman' cultural influences would completely neglect pre-existing local traditions related to water. Instead, it is more realistic to see the hybrid mixing of local and incoming associations, leading to variable interpretations of urban space in the cities of Roman Britain. Some of these local identities might be strong manipulation of local beliefs to the advantage of Roman authorities, but there is also the potential for more parochial investment in water engagement to further long-lived associations with British landscapes.

As is emphasised throughout this book, the physical evidence we have for active Roman water supply in Britain can vary greatly. None of the archaeological evidence presented in this chapter can compare, for instance, to the huge extant structures we can still see on the continent. Moreover, depending on the level of excavation and when it took place, we have relatively sparse remains for even pivotal urban features like the public bathhouses. What follows, then, is a series of case studies that put emphasis on different parts of the urban waterscape. Of course, this approach is partially due to the variability of associations to water in these local settings, as championed throughout this book. However, there are also occasions where the archaeology suggests elements that are yet to be discovered or may have been lost. These may take the form of, for instance, large examples of bathhouses or sewers suggesting an equitable water supply. However, regardless of the evidence, the intention of the following sections is to create greater understanding of how it related to previous interactions with the local waterscape. In doing so, we open up the possibility of meaning beyond traditional interpretations of function.

Lincoln (Lindum Colonia)

Pre-existing connections to water

Lincoln was initially established as a legionary fortress in AD 48, and by around AD 70–80 it had gained the status of *colonia* (Wacher, 1995: 132). There is a perception that the *coloniae* of the provinces were created consciously as archetypal Roman towns (ibid.: 17), and this has been seen as a reason to analyse these settlements in predominantly tactical and economic terms; the more 'Roman', the more likely a modern rationale would fit. In the case of Lincoln, the settlement was originally located on a hill and so was defensible; it was placed close to a geographical bottleneck of the glacial gap in the Jurassic limestone ridge, a vital supply route to the north; and it had links to the river network through the Witham, so could be a centre for trade. These are good practical reasons for the location of such a settlement (and all undoubtedly played a part in the siting of Lincoln), but the landscape of this area cannot be fully explained in practical terms. Indeed, the Roman name of Lindum roughly translates as 'dark pool', which suggests a fundamental symbolic link with the nearby Brayford Pool (Rivet and Smith, 1979: 392–393).

The Witham was steeped in meaning throughout prehistory. The prominent causeway features at Fiskerton, Washingborough, Brigg and Stamp End are examples of the extraordinary reverence afforded to the local waterscape – and Stamp End is also the location of remarkable metalwork deposits like the Witham Shield (Stocker and Everson, 2003). Furthermore, numerous Bronze Age barrows were aligned with these causeway features, showing a long-lived but changing relationship to the Witham in the extended landscape (May, 1976, 1988; Stocker and Everson, 2003: 280). They may also be the source of the name of the local tribe, the Corieltauvi, which has been interpreted as 'the land of the people of many rivers' (Tomlin, 1983; Breeze, 2002). Combine this with the fact that deposits at Fiskerton and surrounding areas continue into the Roman period, and one must start to consider a more nuanced approach to the siting of Lincoln.

The town is located on the banks of the River Witham at a confluence with the Till; here the river channel widens to form Brayford Pool (Figure 3.1). In this body of water evidence has been found of Iron Age occupation on a few sandy islands (Darling and Jones, 1988; Steane *et al.*, 2006). The original Roman fortress (and subsequent *colonia*) was built on a hill overlooking this watery site, but within a few years occupation had extended down towards the Pool. The landscape position, in a glacial gap, meant that establishing a settlement was problematic – the low-lying valley was prone to flooding, and the clay hillside was riddled with springs. Yet in many ways this combination of features meant the site was a nodal point in the wider waterscape: Brayford Pool was not just the joining of the Till

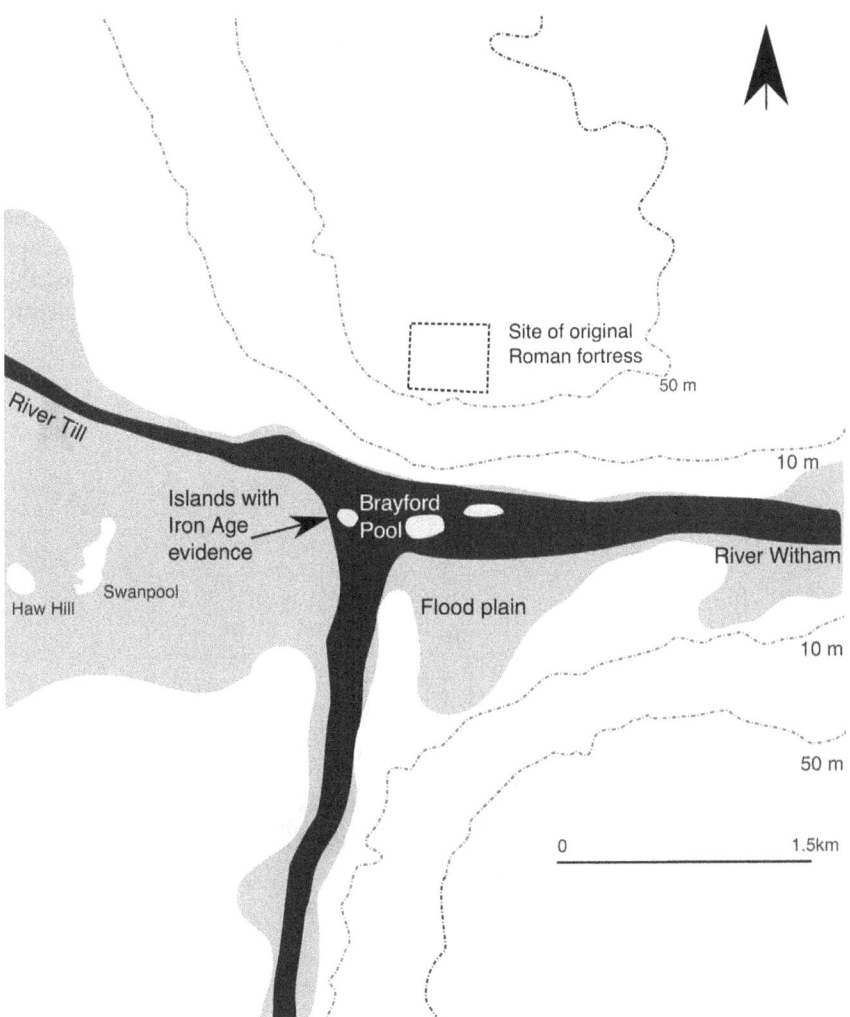

Figure 3.1 The site of Roman Lincoln and its wider waterscape.

Source: Drawn by author after Jones *et al.* (2003a).

and Witham, but also the fusion of spring water from the steep slope with the river water. The Pool was the heart of a more extensive marshy and liminal landscape, all of which could be seen from the high ground of the fortress (Rogers, 2013: 69).

The principal evidence of Iron Age occupation is found on the islands in the centre of Brayford Pool (Jones *et al.*, 2003a: 26). It consists of some roundhouse features that, considering their liminal location, were probably the centre of ritual activity. While this does not amount to a substantial occupation of the landscape, the structures may indicate the broader importance of Lincoln as a meeting place. These dwellings are paralleled by small-scale activity on the nearby hilltop (where the fortress was built). It is possible that movement (or procession?) between the low ground and high ground could have been part of the veneration of the waterscape. During the Roman period a bridge (the Wigford Causeway) was built across these islands, linking both shores of the Witham (Figure 3.2). The tradition mentioned above of ritual causeways in the immediate landscape would suggest that this was far more than just a practical structure for the town. There has been a suggestion that the domination of this ritual site

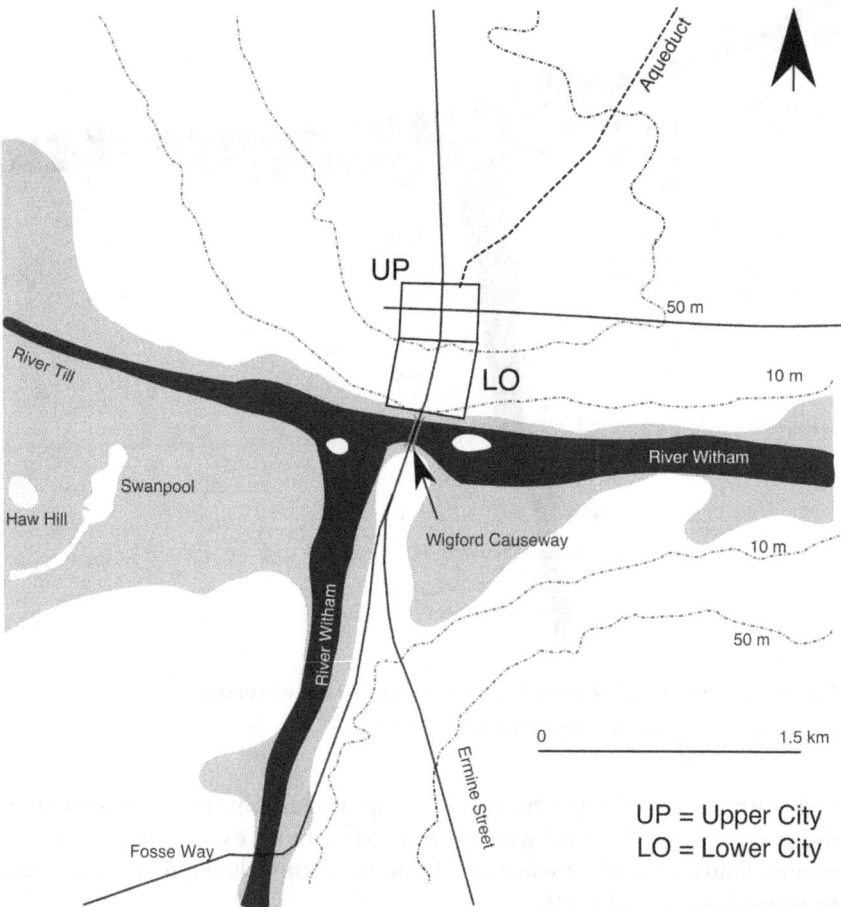

Figure 3.2 The Wigford Causeway and its impact on the waterscape of Lincoln.
Source: Drawn by author after Jones *et al.* (2003a).

within Brayford Pool could be a strong ideological statement made on the part of the colonising Roman force (Jones *et al.*, 2003a: 97); this statement is up for debate.

Due to the explicit military connections, it is easy to envisage the Wigford Causeway as an imposition on a religiously important Iron Age landscape. But as already noted, causeways are symbolically rich features that have a long history of non-practical associations throughout ancient Europe. It is just as possible to see this causeway as an active addition to this meaning-laden waterscape. The bridge would have maintained a sense of liminality, being part of the transition between shorelines. It allowed people to dwell near to and appreciate the complete scope and breadth of the Witham. The recovery from the Wigford excavations of large quantities of drinking-vessel sherds dating to the second and third centuries may also be direct evidence of continued ritual importance for the location (ibid.: 99). Furthermore, the way in which the Lower City at Lincoln expanded towards the waterfront is an interesting part of the equation. While the Upper City was the original site of the *colonia* and contained the forum, the lower part of town eventually featured a number of monumental buildings, potentially including a temple complex, bathhouse and fountain (ibid.: 89) (see Figure 3.3). The bridge forms the first stage in this transition into the central urban space, and ties the liminal element of crossing water into the overarching experience of the town. As discussed previously, there is increased monumentalisation of ritual sites in the Late Iron Age, with the addition of new ways to navigate and experience special places. So this development of Lincoln would not necessarily have been viewed as something alien to the local populace, who continued to change the ways they interfaced with water.

Wells and the Upper City

There is an emerging consensus that much of the future work understanding the transition from an Iron Age settlement to Roman town will come via a more in-depth study of the way the watery environment was treated (Jones *et al.*, 2003a: 54; Rogers, 2013). Water supply must play a prominent role in this analysis. As mentioned, the first Roman imposition on the landscape came in the form of a hilltop legionary fortress in the Neronian period. As with all such military installations, the heart of the settlement was the *principia*; what is intriguing about this building in Lincoln is the incorporation of a well (now known as St Paul's Well). This is not to say that it is unusual for a well to be located in such an area – Johnson (1983: 106), for instance, notes other similar structures in *principia* courtyards. Yet excavation in this area has found no proof that this well was built at the same time as the surrounding fortress. It does not hold a symmetrical position within the *principia* and thus is less likely to

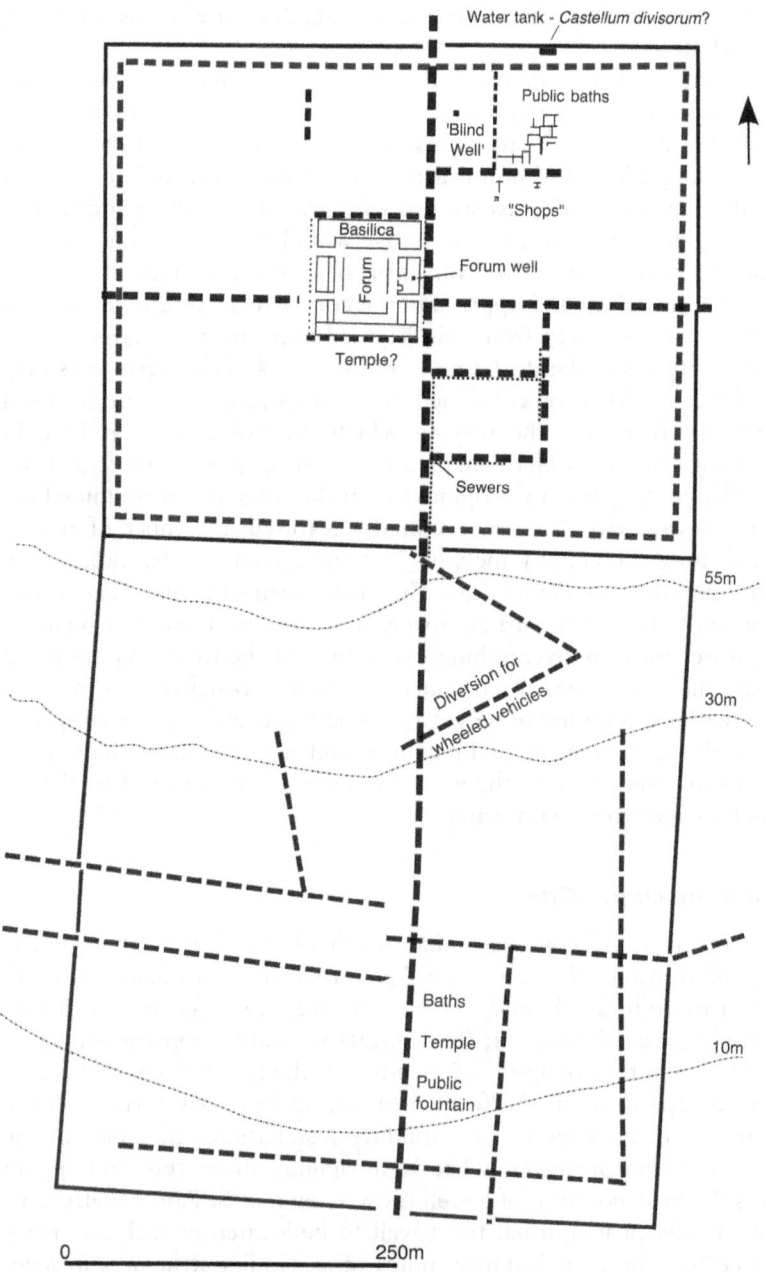

Figure 3.3 Plan of Roman Lincoln with notable features.

Source: Drawn by author after Rowsome (2008) and Rogers (2011a).

be a feature built in the Roman period (Jones *et al.*, 2003a). It was subsequently also incorporated into the forum of the *colonia*, remaining a prominent feature. This enigma has been largely ignored in the overall interpretation of the settlement.

Considering the lack of any eye-catching deposits, it is perhaps not surprising that this well has been given scant analysis. The suggestion that it was an original Iron Age feature incorporated into the Roman-period settlement warrants far more attention, especially considering the aforementioned local significance of water. If we acknowledge the potential ritual significance of the well, it fits established patterns of sanctuaries on hilltops – like Nettleton on the Lincolnshire Wolds (Farley, 2011). Like many sanctuaries in the prehistoric landscape, the Roman forum was built up around this focal point, and hence the monumental statues and structures increasingly framed the experience of drawing water (Steane *et al.*, 2006: 187). The well's importance may also be reflected in its continued use into the medieval period, when it became entangled with Christian meaning and was identified with St Paul (Jones *et al.*, 2003b: 8.1.1). The other known well of Roman Lincoln, the so-called 'Blind Well', was also located in the Upper City and has recently been considered in similar ideological terms (ibid.). This feature was discovered by antiquarian excavations, and was purportedly a massive structure that could well have supplied the bathhouse. It has been considered more recently as being linked to the aqueduct of the town.

Moving away from just isolating the spatial impact of these features, there is also the question of quantifying the relationship between water on the hill and the Witham below. The hill that separates the Upper and Lower Cities at Lincoln is a steep incline which was 'riddled with springs' (Jones, 2003: 111). This high ground gave some commanding views of the Trent Valley to the west and the Witham, plus potential sacred pools, to the south (Jones *et al.*, 2003b). The importance of such a visual link cannot be underplayed. The inclusion of the well in this picture on the high ground gives us a sense of the centrality of Lincoln within a meaning-laden waterscape.

Reinterpreting the piped water of the colonia

Of course, emphasising the symbolic and ritual currency of Lincoln's waterscape should change the way that we interpret Lincoln's impressive aqueduct system. It has been a focus for many archaeologists investigating the city's development (Thompson, 1954; Wacher, 1975; Lewis, 1984; Hodge, 1992). Accounts have invariably concentrated on the potential source of the aqueduct and the technicalities of engineering involved in supplying the Upper City of Lincoln. Interestingly, nearly all these writers discuss the possibility that the aqueduct did not function, but recent archaeological work has meant this is no longer an issue (Williams, 2006). Underlying all this is the assumption that Lincoln was a very Roman town, and this was the principal reason for it possessing the best water supply known in Britain. Only recently have archaeologists begun to question whether assessing this structure in practical terms is limiting. Indeed, Jones (2003) highlights how functionalist research into the purpose of the aqueduct has almost reached its practical end. Furthermore, as described above, the Lincoln area was never really in need of extra water for consumption: its wells and numerous springs would surely have sufficed for this purpose.

Support for a ritual reason behind Lincoln's aqueduct is found in the potential Nemeton (sacred grove) at the Roaring Meg spring. The known trajectory of the aqueduct leads in this direction (Figure 3.4). Furthermore, an inscription found in nearby Nettleham, northeast of the city, mentions the native god Mars Rigonemetos. This was part of a dedication on an arch at the entrance of a temple enclosure (Petch, 1962). The inscription was found in a secondary location but has been interpreted as relating to activity at the spring site. That being said, it is by no means certain that the Roaring Meg spring was the source of the aqueduct; others have postulated sites as distant as Otby Top (Burgers, 2001: 38), a prominent source of water in the second half of the twentieth century, northeast of Market Rasen.

An aqueduct source at Otby Top would link the town to the dramatic natural potential of the Lincolnshire Wolds, which had their own unique resonance in the local community. In particular, it is worth noting how areas such as the Bain Valley are known to be 'storm traps'. A 'cloudburst' recorded on 29 May 1920 famously saw some 11 centimetres of rain falling in less than three hours, claiming 23 lives in Louth alone (Robinson, 2009: 75). Another dramatic account was recorded on 7 October 1960, when a cumulonimbus cloud towering to 12,000 metres lodged against the southwest edge of the Wolds near Horncastle. During the resulting storm 18 centimetres of rain fell in six hours (ibid.: 76). Such events have the power to reshape a landscape and would surely have been explained within a local set of water beliefs. Yet the drainage of these dramatic weather events was focused mostly towards the west (Smith, 2009: 109); the aqueduct would have been the only watercourse leading out of the Wolds towards the east and the settlement at Lincoln. Thus if the aqueduct was sourced from this locale, its flow could have reinforced a link to dramatic weather systems on the Wolds.

A third viable option for the aqueduct source, hitherto not suggested in the literature, is the next valley south of Otby Top. This is the head valley of the River Rase, and it certainly has an appropriate elevation. Moreover, there is a Roman site here above a known spring source at Churn Water Heads (Willis, 2013). The other notable aspect of this site is the fact that

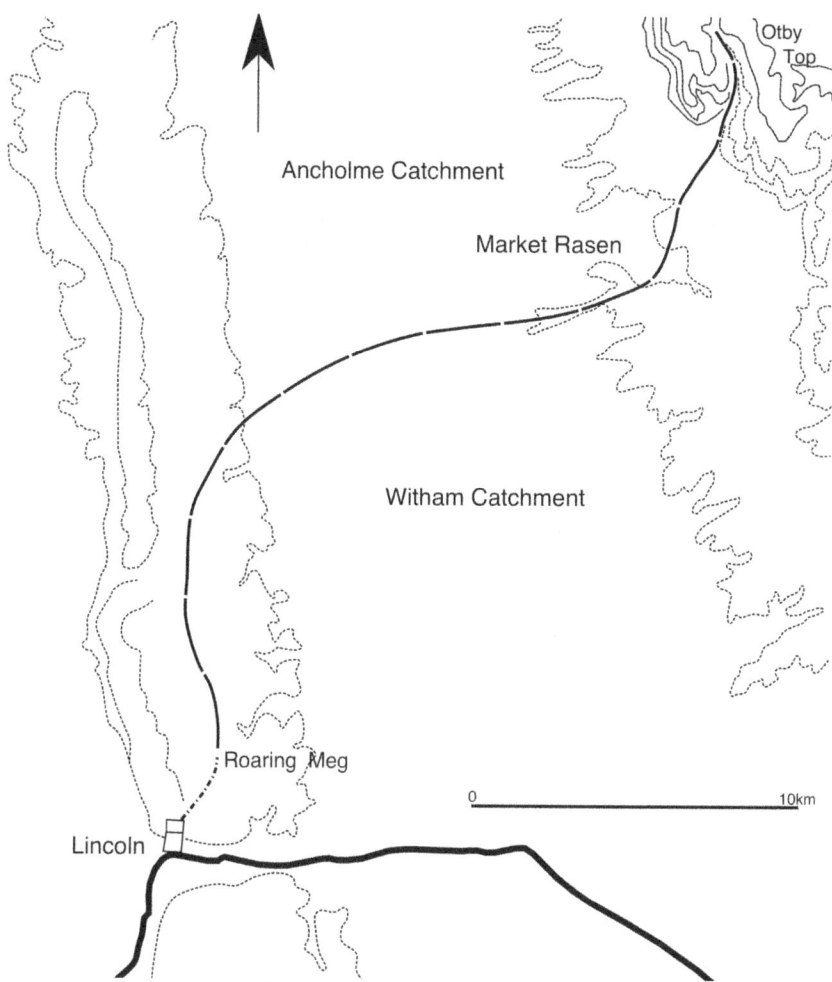

Figure 3.4 The potential course of the Lincoln aqueduct.
Source: Drawn by author after Burgers (2001).

the springs are associated with tufa, so they may have been identified as
being particularly important in the Roman period.

Any of these sources would seemingly fit into the overall schema of
Lincoln proposed in this book. Tracing the aqueduct from one of the
venerated sites in the surrounding landscape to the Upper City once again
reinforces possible mental/visual links between places of watery meaning
and the central areas of the Roman town. It has been shown how both the
bridging of Brayford Pool and the continued prominence of the forum well

could have reinforced local associations about the immediate and wider landscape. This idea of Lincoln being perceived as a nodal point within a highly significant prehistoric waterscape is pivotal when evaluating the aqueduct. It potentially brought new waters directly to the Upper City and created a deeper physical link with the regional waterscape.

Lincoln is also said to have possessed one of the best-developed sewer systems in Britain (Wacher, 1995: 138). Part of this system was uncovered in the nineteenth century, primarily beneath the main north–south street of the Upper City (Jones et al., 2003a: 61). In this area the main conduit was presumably large enough to be cleaned internally and was probably joined by numerous other drains from the east and west (Jones, 2003: 117). Unfortunately there is unlikely to be modern excavation of these features, as many listed buildings are found in the area, but traces of the system have been located close to Steep Hill, meaning it probably ran down to the Lower City and the Witham (Jones et al., 2003a: 61). This appears to reinforce the postulated water-based links between the Upper City and the Lower City. Furthermore, it means that the aqueduct water, probably derived from sacred sources in the urban periphery, ultimately joined Brayford Pool. In essence, this external water could have been seen as creating a new confluence with the Witham.

There have been recent discussions about the potential of the aqueduct feeding water away from the town, perhaps to the sanctuary at Roaring Meg (Jones, 2003; Jones et al., 2003a). It is suggested that this could have been a move by the Roman authorities to appropriate a native sanctuary physically by feeding waters from the forum area to this site. In fact, in many ways this would appear to be an action rather at odds with the more traditional discourses around colonisation. Again, it seems to be an idea based on the waters from the Upper City having meaning imposed upon them by new structures of imperial authority. Of course, if the water was the defining influence, this makes the imposition of a Roman identity on the external site at Roaring Meg somewhat dubious. That notwithstanding, with this interpretation it becomes difficult to justify the size of the sewer system mentioned above and the various water features of the town.

One of the primary beneficiaries of any external water supply would have been the public baths, which were situated in the northeast part of the Upper City overlooking the Brayford Pool area. Despite the full plan of the building never being exposed, remains of several rooms with deep hypocausts and tessellated pavements seem to show this was a building of major significance (Jones et al., 2003a: 79). Indeed, these rooms covered an area of at least 60 metres by 45 metres and show clear evidence of extension after their initial construction. A major rebuilding or modification project looks to have taken place in the late Antonine period or soon after (ibid.: 80). It is possible that the structure had a precedent in the legionary phase of the town, but currently there is only evidence of timber structures within the area. It was noted by Jones *et al.* (ibid.: 79) that, somewhat unusually, the baths do not appear to have had an entrance on the main *cardo* of the Upper City; instead, the east–west road looks to have been the main access point. This does not necessarily have to be a significant attribute, but it could be involved in wider processes concerning the water supply to these baths.

It has been emphasised how the Upper City of Lincoln had a multitude of water sources in the Roman period, and if the aqueduct did siphon water away from the town, it is possible these alternatives supplied the bathhouse. St Paul's Well, with its potential beginnings as an Iron Age focal point and its later role as a central feature of the *principia* and forum, is a highly significant feature. It is close enough to have been a source of water for the baths, and the idea of moving water between the forum and baths could have referenced similar water rites on the hilltop from deep in prehistory. Even from a Roman perspective, the linking of the official power of the forum to another monumental building is symbolically potent. Similarly, the massive 'Blind Well' was located within the same *insula* as the baths (Abell and Chambers, 1971: 19–20). As explained previously, this feature could have been extremely important in Roman Lincoln, but we are limited to antiquarian observations of associated piping (Britton, 1812: 600–601). It could have supplied the bathhouse, and its water was potentially also siphoned out of the town for ritual reasons. Finally, the bathhouse is close to the suspected *castellum aquae*, which is one supposed termination point of the aqueduct. This external water could have symbolically linked the town to the wider importance of the waterscape in this region. We cannot rule out any of these possibilities; in truth, the water supply may have been a combination of all of them. In light of this, the bathhouse would have been the central place where all these powerful waters were mixed and one could interact with them. Moreover, the bathhouse could have fused the associations of these sources before draining the water down the hill to the ritual area of Brayford Pool.

It is also important to recognise the activity in the Lower City. There could well have been another bathhouse in this area, close to the fountain and temple enclosure discussed previously. The discovery of a possible Mithraeum in this zone echoes finds in other towns mentioned in this chapter, perhaps suggesting a connection in Late Antiquity to previous water traditions (see Stocker, 1998). Any bath structure here is unlikely to have had purely practical functions and could well have possessed some connection with the primary building in the Upper City (Jones *et al.*, 2003b: 7.2).

Overall, Lincoln offers certain evidence of the importance of considering a long history of water interactions at Roman sites. There is clear evidence for meaningful human engagement with water deep into prehistory, and the additions of the Roman period appear to be informed by this local history. In fact, it is not an overstatement to suggest that the colony at Lincoln was defined by its relationship to the immediate waterscape – a state of affairs that must lend itself to a hybrid urban identity.

St Albans (Verulamium)

Verlamion and its waterscape

The Roman town of Verulamium (modern St Albans) was pre-dated by an established settlement in the Iron Age. The oppida complex of Verlamion stretched over the valley of the River Ver, with potentially seven major prehistoric occupation sites in the immediate area that would later become the Roman town (Niblett, 1999: 406). The best investigated of these are Gorhambury (Hunn *et al.*, 1990), Prae Wood (Wheeler and Wheeler, 1936) and Folly Lane (Niblett, 1999). According to Rivet and Smith (1979: 498–499), the name Verlamion refers to a 'settlement over [or by] the marsh'. All these sites were at points that overlooked the River Ver and the marshy zone through which it passed (Creighton, 2006: 124) (Figure 3.5). Anthony (1970) notes how the river occupied different water channels over time, and there is evidence for other streams in the area. The location of St Albans was likely a meeting point between the faster-flowing waters of the river and the more stagnant waters of the marshes.

The pre-Roman activity in the vicinity of modern St Michael's, directly adjacent to the marsh and underlying central areas of the Roman town, almost certainly held prime importance in the wider landscape. The concentration of metalworking debris and deposits is unparalleled on other sites in Verlamion (Niblett, 1999: 411). Furthermore, nearby evidence has been found for a crossing point of the Ver. Three rows of timber uprights have, somewhat misleadingly, been named the Timber Tower, but are generally interpreted as part of a brushwood trackway, perhaps in association with a gate (ibid.: 411; Niblett, 2005: 65). A number of what are thought to be votive items were found in the river peat close to the crossing, including a bronze patella and a late pre-Roman Iron Age ceramic bowl found below the posts of the structure (Niblett, 1999: 411; Rogers, 2013: 31). There were also numerous coin moulds in the area that could have a ritual explanation (Niblett, 2001: 61). The recovery of a small plaque depicting a rudimentary but Roman-style river deity lends credence to the interpretation of the Ver as ritually significant beyond prehistory (ibid.: 87). This causeway across the marsh led up to the Folly Lane temple complex, which was a pivotal site in the local Iron Age landscape and of continued importance into the Roman period. Indeed, Creighton (2006: 127) details how the temple complex was directly and purposefully linked to the St Michael's enclosure as a processional route. The Roman forum-basilica was eventually built over this enclosure, and thus maintained the prominence of a link

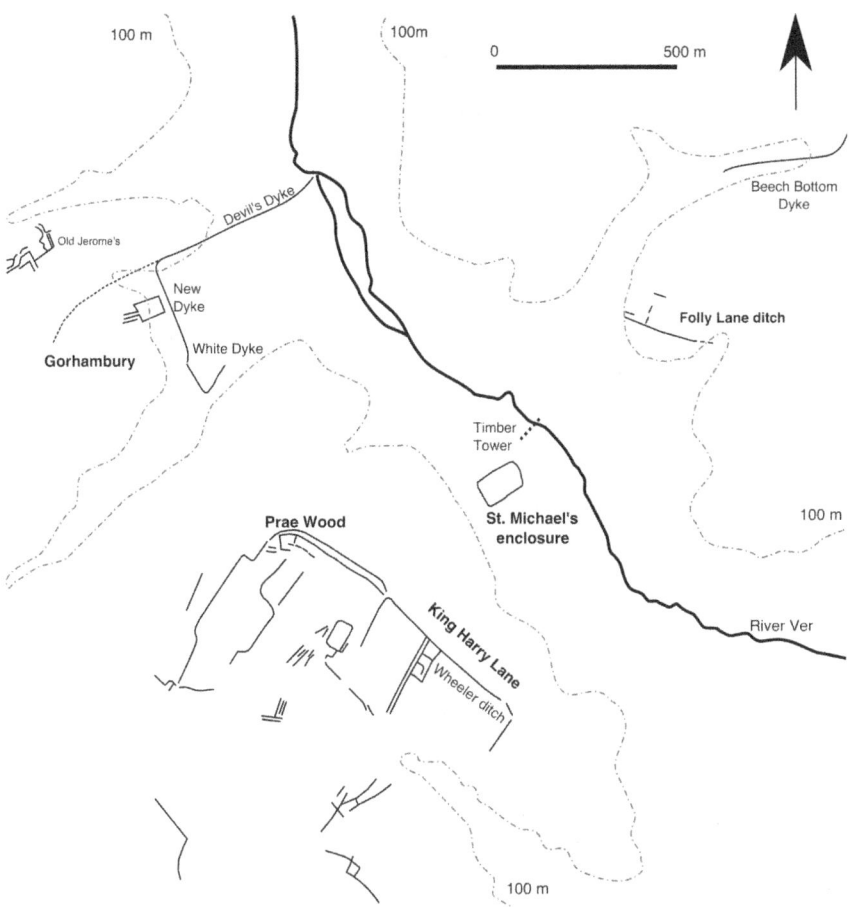

Figure 3.5 The pre-Roman landscape of Verlamion.

Source: Drawn by author after Niblett (2001).

to Folly Lane. Creighton's discussion emphasises that the continued importance of Folly Lane dictates the monumental urban landscape of the Roman
period; the bathhouses, forum and theatre all seemingly address the land-
mark, and were likely part of the processional route towards it (ibid.).

Much of the discussion of Verulamium's prehistoric context focuses on
the central role of the Folly Lane complex, rather than seeing it as an
aspect of the broader importance of its landscape context. However, the
engagement with water at Verlamion and Verulamium is central to any
connection between the central St Michael's enclosure and the site at Folly
Lane. As mentioned, the act of crossing the Ver via causeway or bridge

appears to have been symbolically profound, and in fact this could be the real foundation for ritual activities in the area rather than the site at Folly Lane. At the start of this chapter there is a brief discussion on the import-ance of monumental earthworks and their symbolic connections to water. At St Albans there were some prominent examples, including the Devil's Dyke, New Dyke and Beech Bottom Dyke, which were all within the same area relatively close to the River Ver (see Figure 3.5). The latter of these was a massive construction of indeterminate purpose. Bryant (2007: 72) has proposed that rather than being merely a boundary, it could have been part of a more extensive processional route, with the earthwork flowing into the valley of the River Ver, and marking out the river crossing and the Iron Age enclosure. The similar Devil's Dyke may have symbolically connected the Rivers Ver and Lea. The White Dyke may even have had a physical connection to the waterscape; spanning 23 metres across bank and ditch, it harbours evidence of molluscs, suggesting it was once filled with water (Niblett, 2005: 31). Hence the Folly Lane temple complex could well be a further engagement addressing this waterscape, rather than a principal instigator of developments. This, in turn, should make us ques-tion events in the Roman town, in particular the structural engagements with water.

Branch Road and Folly Lane excavations

In this vein, it is interesting that in British terms St Albans has a great deal of evidence for water supply and management. One can start by noting that in the intermediary landscape between Folly Lane and the central Roman town there were 28 'shaft' features found southwest of the ceremo-nial enclosure (Niblett, 1999: 83) (Figure 3.6). They date from around the middle of the second century, so they coincided with a period of intense building/rebuilding in Verulamium. At least eight of these are believed to be wells, but the lack of complete excavation means the distinction between a 'shaft' and a 'well' is not entirely clear. While the resources were not in place to excavate the 'well' features fully, their potential depth was much more than that of the surrounding pits, which were around 2 to 3 metres (West, 2012). Fulford (2001) notes that many of the shallower fea-tures would not have pierced the water table; but there is potential that these pits could have accumulated rainwater (maybe via the natural run-off of the slope) and acted as a type of 'dumb-well' (West, 2012). Indeed, as in the case of the pit alignments mentioned in Lincoln, accumulation of water in this type of recess could have an ideological/ritual element, even if it did not reach the water table. It may even be possible that water from the deeper 'well' features was procured and stored temporarily in the sur-rounding shallow pits. This could facilitate greater interaction with the water for any associated rites.

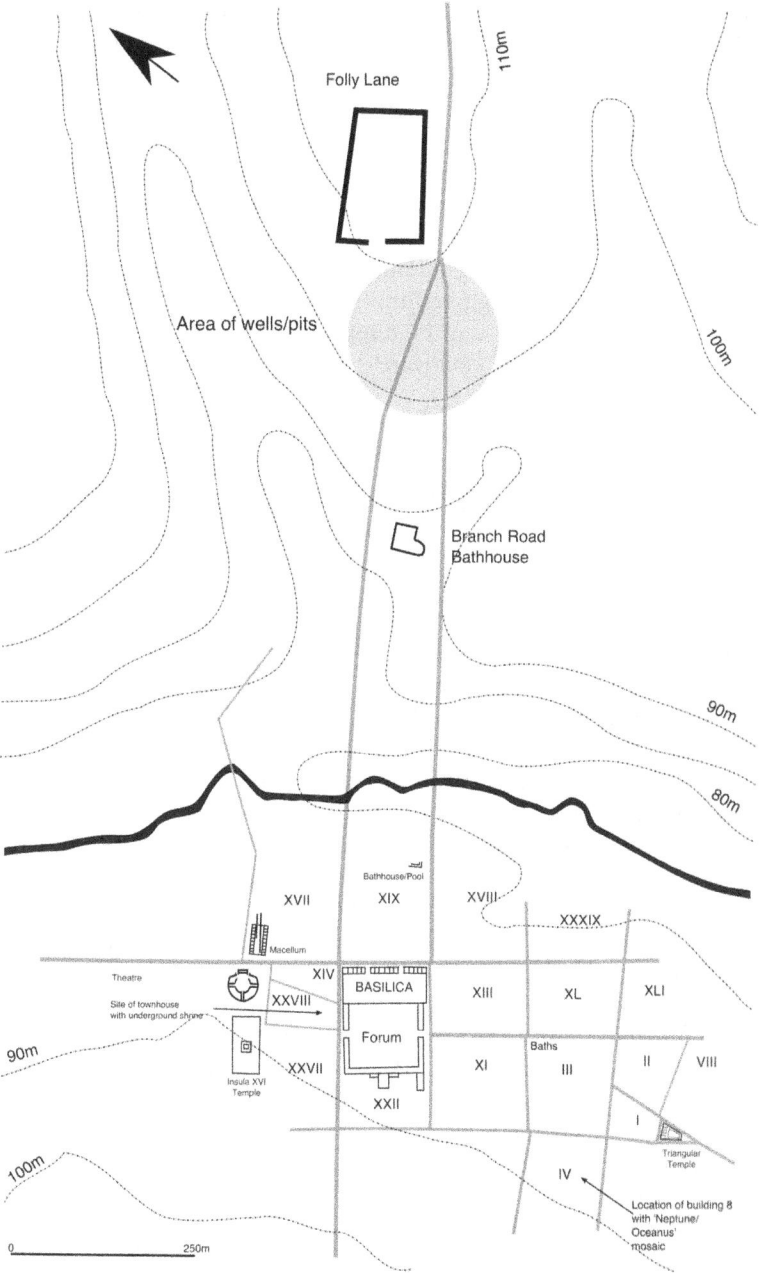

Figure 3.6 Roman Verulamium and the Folly Lane temple complex.
Source: Drawn by author after Niblett (2005).

These features seem to be characterised by votive deposits, such as human and dog skulls, plus two massive deposits representing at least 34 cattle (Fulford, 2001: 210). A number of the shafts also contained complete or partially complete pots in both the primary fills and the sinkages above (Lyne, 1999: 299–302). While these features may be of a second-century date, the human skull in particular does echo the importance of the head in rituals of the temperate European tradition. Interestingly, seven of the pits contained the remains of face pots. It is possible that these are symbolic representations being deposited instead of skulls, thus marking a form of continued local belief. That notwithstanding, the proximity of these features to Folly Lane suggests another aspect of water interaction involved in the movement between the temple enclosure and central areas of the town. Moreover, like the enclosure itself, this activity does not seem entirely 'Roman' or 'native'.

Not far from these wells, on the road approaching the town from the direction of the Folly Lane ritual complex, is the Branch Road Bathhouse. The positioning of this water focal point, in this intermediary zone between Folly Lane and the main town, is not coincidental. Pottery from the baths dates its construction to circa AD 140, meaning it was being built at a similar time to the refurbishment of the complex at Folly Lane (ibid.: 290–291). It also faced the ceremonial site, like some of the other structures in the town. Its dimensions were at least 55 metres × 33 metres, but the limits of the excavation mean that a significant portion of the building remains undocumented. Niblett (2005: 84) notes that the north-east and northwest sides of the building could even have stretched towards the Verulam Road; quarrying in the nineteenth century would undoubtedly have affected much of the nearby evidence. The site seems to have been in use until the second quarter of the third century, when water-borne silts were accumulating in the hypocausts (ibid.: 85). While the baths possessed rooms that one could interpret as a *frigidarium*, a *tepidarium*, a *caldarium*, etc., there is little evidence of regular use that would befit a pragmatic roadside bathhouse. This lends weight to the interpretation that the structure was used for seasonal festivals and rites emanating from the Folly Lane focal point (ibid.: 84). One then has to wonder whether the progressive flooding of the structure could have been ceremonial, or an open acknowledgement of the power water held in the immediate surroundings. Furthermore, in a changed state it could still have been utilised for many different activities – at the very least it may have been a focal point before crossing the Ver into the town.

The water supply for this bathhouse is somewhat conjectural at this point. It is possible that there was a stream running from the high ground of the Folly Lane area, and this could be the reason why water eventually amassed in the structure. Large masonry features were discovered in an excavation at 51 Folly Lane, and could be evidence of an aqueduct that brought water to this external area (Niblett and Thompson, 2005: 87). This would be of some note, considering this bathhouse is outside the inner areas of the town and thus would be the exclusive focal point for this water. It would certainly lend credence to the notion that this was a special building, potentially connected to the significance of the important local waterscape.

Water in the central insulae of the town

Regarding water supply to the central areas of Verulamium, aerial photography gives evidence of a presumed conduit leading towards the town from further upstream on the Ver. A shadow seems to be heading towards the Chester Gate area of the town, but the exact location of its termination is unclear. As there has been little in the way of sustained fieldwork on the feature it is difficult to estimate its date of construction, but it is possible that it was a second-century development (Niblett, 2001: 115). This suspected aqueduct appears to leave the Ver in the area of Redbourn, which is not insignificant considering the presence of a known Iron Age hillfort nearby; it is also not too distant from the Friar's Wash temple complex that Time Team discovered in the 2000s (Wessex Archaeology, 2009). Moreover, the proximity of this complex to the Ver suggested that it played a principal role in any religious rites taking place (ibid.). Thus it is of significance that an aqueduct could be siphoning this water down to Verulamium.

The problem for any aqueduct entering Verulamium from this direction at the Chester Gate would have been the relatively low elevation in comparison with some areas of the town (Niblett, 2005: 87). It may be that it was supplemented by springs on the high ground, with associated small streams also playing a part. For example, there is evidence in the late pre-Roman Iron Age of a small valley, now almost entirely silted up, running from the region of the south corner of the Roman forum down to the river (ibid.). A perched water table or streamlet fed by springs could have run along this route, and may have been sufficient to provide water for specific features or buildings in the areas of higher elevation. The discovery of timber pipes from Insulae II, XVIII and XXVIII suggest there was a significant distribution of water in the town by the end of the first century (Niblett, 2001: 77).

There are also rather interesting uses of this water in the monumental heart of the town, with Insulae XVII, XIX, XIV and XXVIII all possessing a mix of water features and ritual foci (Figure 3.6). For instance, a mortar-lined conduit, which appeared to bring water from the Ver to a point opposite the theatre, was discovered in Insula XVII (Niblett and Thompson, 2005: 86). Close to this point of termination, a fragment of a sizeable Purbeck marble basin was recovered from the plough soil; it has been interpreted as possibly being derived from a fountain.[1] However, this conduit also led directly through the so-called *macellum*, which has caused many to question the definite purpose of the building. In three stages of development the building became a monumental structure, with a central nave primarily taken up by masonry bases for two large tanks. The provision of high-volume water supply and the impressive size of the structure have cast some doubt on its definition as a *macellum*.

There are not many definite examples of *macella* found in Britain (Rogers, 2008: 155). Closer to the heart of the Empire they were a staple form of building present on many different sites; they usually possess a pattern of small rooms bordering a generous open space (de Ruyt, 1983). Yet the British examples that have been identified do not necessarily have a uniform design. Regardless of this, there is a sweeping popular conception of them as being just markets. Undoubtedly goods were bought and sold in these areas, and thus part of their role would have been similar to the market of today. However, as mentioned in Chapter 1, aspects of rituals involving the sale of food were commonplace in the Mediterranean. Combine this with its communal nature, which inevitably brought together a diverse swathe of the population, and one is presented with a dynamic space that could have been conceived in various ways. It has been noted that applying our knowledge of Mediterranean *macella* to the periphery of the Empire can be somewhat troubling (Ellis, 2000: 342). To this end, the rather elaborate water supply and drainage features in these buildings are of some interest. Niblett (2005: 105) postulates that the structure at Verulamium could have been a *nymphaeum*. But it is possible to see both interpretations validated in the long life of the structure. It appears rather short-sighted to ignore any local water rituals in deference to a Roman construct, particularly when there is also evidence in the same insula for two Romano-Celtic temples (ibid.: 91).

Another building close to this *macellum* structure was the town's theatre. Creighton (2006: 127) notes how the orientation of this theatre was towards the Folly Lane enclosure on the other side of the Ver, and was likely a point of communal activities related to this ritualised processional route. As the Ver was the source of a conduit supplying the *macellum* area, the water used in the central monumental features of the town may have furthered this relationship with the external sanctuary and its watery context. There are also a number of Antonine vaulted drains adding to this picture. One ran from the southwest corner of the forum to the river (Frere and Wilson, 1983: 58–59); it had practical uses, such as flushing a nearby latrine, but that does preclude a more symbolic meaning. Another pair of similar features interacted with other significant monumental buildings: one drained the orchestra of the theatre, and the other appears to have emptied into a ditch close to the Chester Gate (Niblett, 2005: 90).

Closer to the river, excavations in Insula XIX undertaken by Saunders (1974, 1975) uncovered a number of pits, fragments of pre-Roman coin moulds and a potential Roman bathhouse. The remains consist of a substantial mortared flint building resting on chalk, flint and tufa footings. Two parallel walls were also excavated – one of which had internal buttresses and an elaborate painted design (Niblett, 2005: 85). Flue tiles were recovered from within the building, in addition to a sunken mortar-lined area that could have been a pool feature. All this suggests either a bathhouse or another structure of some stature with a central watery focus. The flue tiles were of a pre-Flavian type, meaning that the structure could have been of a relatively early date. The evidence of burnt timber could indicate that the building was established prior to the Boudican revolt of c.AD 60–61 (Niblett, 2001: 65). However, even if this were the case the event did not seem to affect the use of the structure, which remained until at least the end of the first century (Niblett, 2005: 65). As mentioned, this area of the town fronted on to the River Ver, and was also closest to the Timber Tower and the prehistoric crossing point. Moreover, the aforementioned bronze river deity was recovered in this area (Niblett, 2001: 87), so this ambiguous building may have been important in the same vein as the Branch Road structure.

Similarly curious in the areas of the town close to the Ver is building 1 of Insula XXVIII. The structure seems to have been a centre for piped water at different points in its existence, and possessed a substantial drain (Frere and Wilson, 1983: 242). In addition, it had an intriguing underground corridor complete with an lapsed niche; this provided a focus at the end of the 13 metres long passage that was large enough to house a life-sized statue (ibid.: 249). It appears entirely possible that this was an underground shrine, although it has no remains that directly imply either Mithraism or Christianity (ibid.: 250). While we lack any evidence to add to the interpretation of this subterranean feature, it was accessed via a sloping ramp down from the street. This road would surely have been one of the busiest in the town, en route to the forum and also leading to the original crossing point by the Timber Tower, resulting in a direct link between this putative underground shrine and the more extensive processional navigation of this landscape discussed earlier.

Finally, Insula XIV is worth mentioning. A deep pit in this area (pit 7) produced contents that were consistently datable to the Flavian period (Frere, 1972: 23). The depth of the feature led the excavators to interpret it as a possible well. The samian vessels deposited within the pit had wear on the foot-rings, meaning they did not derive from a pottery shop but were potentially used in communal rituals (Woodward and Woodward, 2004: 81). Crucibles were found, which could link the feature to metalworking and some of the ritualistic associations afforded to that activity (Budd and Taylor, 1995). There was also an impressive series of major fragments from glass vessels (Charlesworth, 1972). This was the only pit that was fully excavated; further work could have established some of the other nearby features as wells.

Southeastern areas of the town

Insula III has had relatively little in the way of archaeological investigation. However, the footings for a substantial masonry building containing a hypocaust were recovered in 1998, located in a single trial trench cut across the line of Watling Street (Niblett, 2005: 85). These were contemporaneous with a series of large masonry drains leading north (ibid.). The scale of both these discoveries means the building was unlikely to have been a private residence; rather, it is probable that this was the location of a primary public bathhouse in the town, dating from the late first or early second century (ibid.: 86). These baths could have been in use until the end of the third century, before the building was either rebuilt or fell out of use.

The position of these baths is not as intimately related to the River Ver as the previous examples, but there is some evidence to suggest that water in this part of the town was still involved in ritual and symbolic activities. The first aspect to note is the proximity of this building to the wealthy residence in Insula IV that contains the famous Neptune mosaic. The baths included a notable well feature, located in a projecting room on the northwestern corner which was entered at the eastern side of the room from the adjacent veranda. The total depth of the well was around 11.5 metres, which, according to Wheeler and Wheeler (1936: 104), would have been insufficient to gain access to water. Indeed, the Wheelers assessed that this well was never actually in working order, despite being competently built (ibid.). They backed this interpretation up with an analysis of the fill; instead of consisting of a gradual accumulation of material, it seemed to reflect mass filling of the shaft at one time. The dating evidence stems from three coins of Antoninus Pius, plus a selection of pottery attributed to Antonine potters from various levels.[2] Willis (2005: 12.3) highlights the prominence of samian ware within the fill and how this mirrors similar well deposits from a range of sites, all of which have been interpreted as examples of ritualistic behaviour.[3] The overall pottery sample contains a variety of items, but generally they are drinking vessels and open forms; there is thus a sense that feasting activity could be a part of the life (or death) of the well. Another corroborating factor for this interpretation is the abundant food debris, including bones of an ox, sheep, pig and indeterminate birds, large quantities of oyster, mussel and whelk shells, and lobster claws. Among this assemblage, the seafood evidence seems striking. The lobster claws, in particular, are surely intriguing in any well context; but with the proximity to the impressive mosaic depicting Neptune, complete with customary lobster claws,[4] a ritual interpretation seems inescapable (Niblett and Thompson, 2005: 89).

The unusual deposits in the well, combined with the mosaic, suggest that structural engagements with groundwater were given respect. Even though this well probably did not source a good supply, beliefs involving these actions perhaps resulted in more reverence – the lack of water may have been viewed as the gods taking offence at this incursion. The surrounding residences of Insula IV may also have had private bathhouses of some significance. This would make this area of the town synonymous with water-bearing structures, and the public baths mentioned above may have been the central feature of a water-focused district.

In line with this, it is worth noting how Insula III (and its public bathhouse) lies at the end of the unusual diagonal course of Watling Street after it enters Verulamium. Along the route of this road is the triangular temple, which is thought to have been a place of worship before the Roman period; a collection of 18 pre-Roman and early Roman brooches were discovered under the southern end of the temple (Wheeler and Wheeler, 1936: 113–120). Taking this into account, perhaps the route from this site towards the central areas of the Roman town (and the prehistoric activity on the Ver) is part of the reason for the diagonal construction of Watling Street. The termination point at Insula III could have held meaning in terms of a processional route, in a similar way to the interpretation of Folly Lane. Within this framework, the establishment of a monumental bathhouse at this point could have underlined long-lived ritual movement through the landscape, rather than merely the increased Romanised values of the town's inhabitants.

Water and St Albans

The work of Niblett (2005), Creighton (2006) and Rogers (2013) has done much to outline the clear connections between the ritualised prehistoric landscape of St Albans and its Roman-period development. The processional pathway between the Folly Lane site and the central areas of the town is now an accepted part of the geography of the settlement, and a defining concept upon which we evaluate the contraction of primary buildings in the Roman period. But what has been underemphasised is the extent to which engagement with water proliferated in all aspects along this route between the two prominent sites. Bathhouses, aqueducts, sewers and wells, with reasonable evidence to suggest they are more than merely practical developments, are central to how people would have experienced the town. The prehistoric activity appears to involve increasing engagement with water, hence Roman-period developments like aqueducts and bathhouses should be seen as playing a part in such long-lived processes, albeit in more monumental and altered form.

The relationship between Verulamium and water is so pervasive that it continues to define the settlement in later periods. Esmonde Cleary (2005) refers to the continued importance of the Ver in later Christian accounts. In the *pasio* of Alban there is a direct reference to people thronging to the bridge as the future saint approaches. To reach the other side, Alban walks upon the water; this is a crucial part of his eventual beatification, as it represents a miracle but also marks the journey to his death in the nearby arena. The Christian habit of appropriating important symbolic concepts and local myths means such a reference cannot be discounted. Alban's crossing is a miracle hinting at a tradition from antiquity. We could even postulate that this alternative route Alban uses may be close to the original crossing of the Timber Tower and the processional routeway that was so connected to the local waterscape.

London (Londinium)

London – a Roman waterscape?

Underlying modern London, the Roman settlement of Londinium was sited in a landscape fundamentally defined by water. We often emphasise the influence of the Thames, but this primary river was joined by a number of smaller watercourses – including rivers like the Fleet and Walbrook. One must note how the area of Southwark was formerly composed of many low-lying islands. Tidal and broader (at over 300 metres during high tide) than it is today, the Thames at this locality would have been a striking setting. It undoubtedly had practical advantages, being a low point for bridges and a clear navigable trade route out to the continent. Of course, the fact that it was the most suitable bridging point means we are discussing a place that was intrinsically different and special in the wider waterscape. The same physical characteristics that make this point important in a prac-tical sense could lead to differentiation in the wider understanding of the place. We should also acknowledge that while the permanent bridge across the Thames was a new development of the Roman period, less permanent forms of crossing had been part of life for people for many generations pre-viously. This means that knowledge of the river was probably well developed and any points conducive to crossing would have been identified and utilised, resulting in many layers of cultural experience. Regardless, this practical perspective has overwhelmingly defined discussions of London's Roman waterscape. Only recently has Rogers (2008, 2013) gone some way in counteracting this overarching logic, proposing the nuanced relationship the settlement had with water from prehistory through to Late Antiquity.

It is often said that Londinium was a new site built for commercial pur-poses, with no origin in the Iron Age (Todd, 1989: 79; Perring, 1991: 1; Rowsome, 1998: 35). This is somewhat misleading: while no recognisable 'town-like' settlement existed, there was activity on the site dating back to the Bronze Age. Also, by the first century AD this location on the Thames was seemingly a point on the border of around five tribal territories; thus its meaning does not necessarily need to be quantified in terms of a defined settlement. The clearest archaeological evidence from prehistory is in the form of structured deposits relating to the water of the Thames and Walbrook, and the marshy islands of Southwark. As mentioned in Chapter 1, Fitzpatrick (1984) studied the numerous prehistoric deposits recovered from the Thames and highlighted their potential ritual significance. The unusually high number of human skulls (Bradley and Gordon, 1988) discovered in the context of London's rivers has been a point of debate. The fact that many are from young males and appear to have been significantly altered (mandibles removed) suggests prehistoric ritual activity (ibid.; Marsh and West, 1981).[5]

The island area of Southwark harbours some evidence of ceremonial activity in the late Bronze Age and Iron Age. For instance, close to the later Roman bridging point a ring ditch was discovered in association with the cremated remains of at least eight children or juveniles (Heard et al., 1990: 610; Brigham, 2001: 10). There is little evidence of in situ burning, but the numerous spreads of charcoal and cremated bone suggest nearby pyres (Brigham, 2001: 10). An area of 'compact silty loam' may represent the last remains of a mound covering this central feature. Later Iron Age evidence is also apparent, with an unusual inhumation burial found at 124 Borough High Street (Heard et al., 1990: 610). The position of the body, with legs drawn apart and head raised, seems to be carefully structured. Unusual positioning of the human body in burials has often been seen as evidence of heightened significance (Lambot, 1998). Rogers (2008: 95) rightly notes the extent to which the massive later activity in London may have truncated the evidence of prehistory at the site. Nonetheless, even if we cannot see a considerable *oppidum*-style settlement, there could still have been more ephemeral ritual significance centred on aspects of the waterscape. What certainly becomes clear below is the extensive role of water within Roman London: it played a key role in almost all aspects of the monumental buildings, and likely had a huge impact on the way one moved through the settlement.

Southwark and the Thames waterfront

The potential for local ritual and symbolic meaning for Southwark, for instance, seems to be borne out in the Roman period. Perhaps this should not be surprising: the marshy island character of Southwark is a waterscape ripe for such associations. Rogers (2013: 44–45) wrote about the extensive river revetments that were needed to maintain the viability of this landscape in the Roman period. As noted previously in this book, such human engagement with water appears to be symbolically productive and profound in the ancient world.

In line with this, at Swan Street there is a group of 15 wells, at least 6 of which contained damaged ceramic vessels and other evidence of ritual activity (Beasley, 2006). These date from the first century right through to the third century, displaying a long-lived interest in ritually charged engagements with the water of the area. This type of early activity was subsequently given monumental form with the establishment of the sizeable religious precinct immediately adjacent to the southern island of Southwark, at Tabard Square (Durrani, 2004). In this area there is evidence of two Romano-British temples dating from the second century (Killock *et al.*, 2015) (Figure 3.7). Esmonde Cleary (2015) and Killock *et al.* (2015: 256–257) both suggest the major temple precinct could have had a direct connection to water worship. The dedication recovered from a Gaulish *moritix* (shipper), the proximity to water routes and the setting in a lowland marshy landscape have all been put forward as possible links. There has also been an assertion that the Tabard Square complex was a nodal site that could have connected many notable rural temple complexes on the routes into London. The sequence of sites along Watling Street – Springhead, Greenwich Park and then Tabard Square – is particularly notable in this regard (Killock *et al.*, 2015: 257). Springhead is one of the best examples of a water-based rural sanctuary in Roman Britain (Andrews *et al.*, 2011), and its alignment with and similar development time scale (in terms of its expansion in the second century) to that of Tabard Square are potentially revealing.

Closer to the Thames there is further evidence of connections between Southwark and religious activity. For instance, there was a potential religious function for the monumental structures at Winchester Palace (Rogers, 2011b: 213). This is supplemented by interesting finds, including a marble figure of Neptune and limestone figures of a genius and a native hunter god (Merrifield, 1983: 188). The figure of Neptune is relatively early in date, possibly being a first-century sculpture. In addition, Merrifield (1987: 49) highlighted the special nature of two late-second-century pits and a well in the area (F28–30), due to their unusual deposit contents. Following in this vein, Seeley and Wardle (2009) explored the environmental evidence from a variety of wells in Roman Southwark. The fill of these structures contained many unusual animal bone assemblages; the deposition of dogs, antlers, chickens and exceptional pottery vessels together seems to point towards a ritual interpretation. In many ways this reinforces the importance that this set of islands held in the Roman period. While some of the most remarkable finds have come from later third- and fourth-century fills, these may be final acts of respect to a meaning-laden structure that was first dug some years earlier. Moreover, we could also be seeing a physical manifestation of a previously more ephemeral acknowledgement of importance. Either way, the votive character of deposits in Southwark is a clear and unique aspect of the local experience of these islands, and they say something about the special way this area was viewed in the Roman period.

Generally these wells occur close to the main thoroughfares that lead to the Thames bridge crossing – as can be seen in Figure 3.7, which highlights

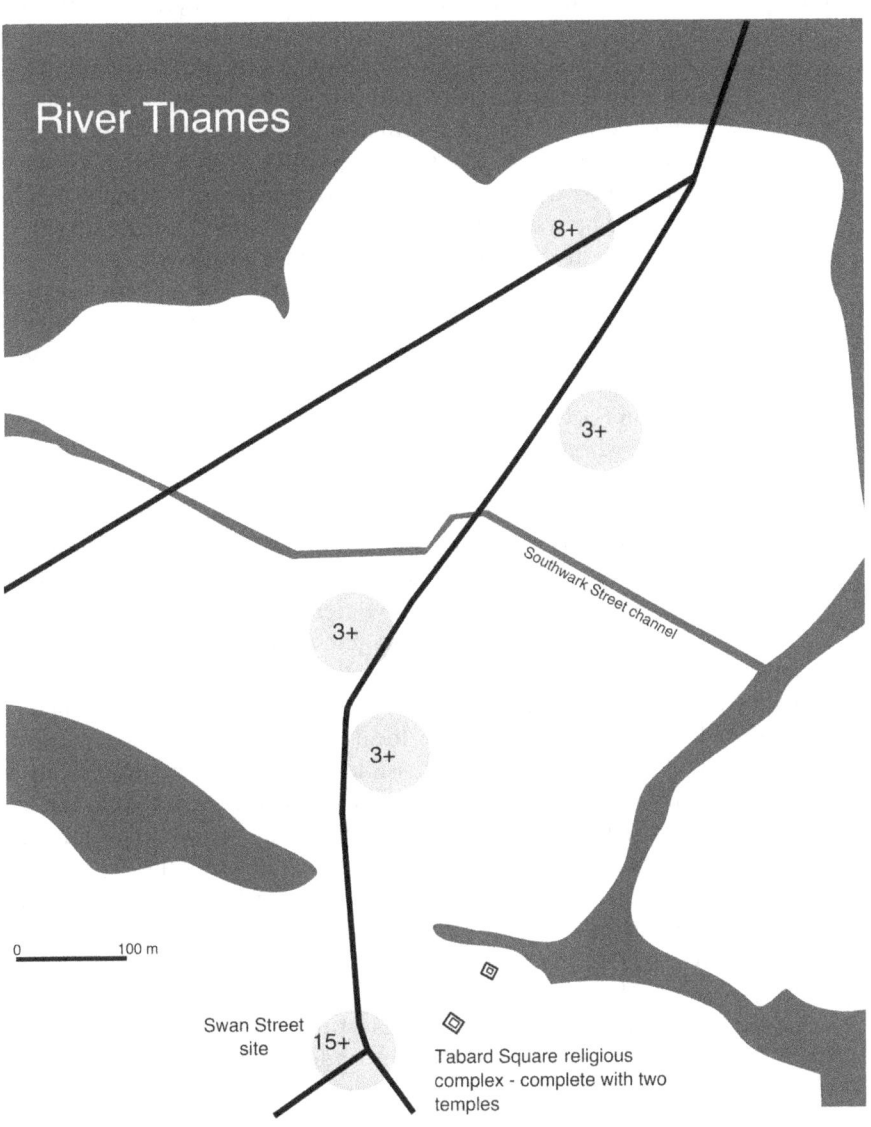

River Thames

8+

3+

Southwark Street channel

3+

3+

0 _____ 100 m

Swan Street
site   15+

Tabard Square religious
complex - complete with two
temples

Figure 3.7 Southwark, the Tabard Square temples and locations with notable
     well concentrations.

Source: Drawn by author after Rowsome *et al.* (2011).

the areas with the highest concentration of wells (Cowan, 2009: 29). They could therefore feasibly be another manifestation of the important connections between water and routes into London, continuing the connections outlined above with temples on the road to Tabard Square. The bridge, of course, was likely one of the most remarkable interactions with water in Roman London (Brigham, 2001: 30). The early evidence for this ancient crossing was found as a result of the nineteenth-century London Bridge rebuilding project. The work of Smith (1841) to retrieve and catalogue the evidence dredged up in this building process has left us with a significant record of archaeological material associated with the ancient crossing. Most of the work regarding this structure centres on the practicalities of construction (Milne, 1985), but Smith (1841) found a series of wooden piles and a concentration of Roman coins associated with the apparent supporting structures of the bridge. Rhodes (1991: 184) notes that this was probably evidence of the ritual activity taking place on the bridge, suggesting a shrine as the reason for the coin deposits.

Rogers (2011b, 2013) convincingly argues how the waterfront structures surrounding the bridge were also far more than merely practical constructions. The timber quay at the St Magnus House/New Fresh Wharf site was a monumental construction with five tiers of large oak beams held in position by a framework of braces and piles (Miller *et al.*, 1986: 8). Of course, such a robust structure was necessary to accommodate the commercial traffic of a settlement on the scale of Roman London, but negotiating this deep interaction with the local waterscape could have been a complex dimension of the more visible economic activities (Rogers, 2011b). This idea of sanctioning such activity could be reflected in the find of over 400 unused samian vessels, including cups, bowls and lion-head *mortaria*, in the fill of the timber quay (Miller *et al.*, 1986: 49).

Some 300 metres north of the bridge the road continues straight to the Roman forum and basilica. These constructions were colossal in scale: the second incarnation of the basilica was 52.5 metres wide and 167 metres long (Marsden, 1987: 38). The building work involved the deliberate dumping of around 20,000 cubic metres of materials to elevate the structure (ibid.: 39). This was a gargantuan undertaking, and has been portrayed as an effort to dominate the surrounding landscape. Creighton (2006: 106), for instance, notes how intervisibility between the forum and the bridge could have emphasised official processions, with a ceremonial crossing symbolically reconquering the territory. However, it seems plausible to suggest such a procession could be less about aggressive domination of the landscape and more related to an exhibition of legitimacy in a meaning-laden landscape. Perhaps aspects of Southwark's religious connections, such as the Tabard Square temple complex and the votive wells, played a role in such ceremonial movement through the town.

Defining space: rivers in Central London

The Thames and its bridge are just one part of a dynamic waterscape of the settlement. The Walbrook was far more integrated into the central areas of Roman London and the supply of water to central structures. As mentioned, it is a watercourse associated with a high volume of skull finds with a broad date range, possibly stemming back to the Bronze Age (Rogers, 2011a: 70) (Figure 3.8). The special nature of this watercourse seems to endure throughout the Roman period. The third-century *mithraeum* (Shepherd, 1998), which now dominates discussion of religious space in this area of Roman London, is thought to have been one of a cluster of temples in the Middle Walbrook Valley (near modern Bucklersbury House); this activity denotes a religious presence up to Late Antiquity. A sculpture of a river god was recovered from this area; according to Toynbee (1962: Plate 35, No. 29), the style suggests it was made in the reign of Antoninus Pius or Hadrian. Charred remains of arcaded timber panelling and a face urn were also found in the area (Wilmott, 1991). Henig (1998: 232) suggests a shrine of the Dioscuri, and Bird (1996)

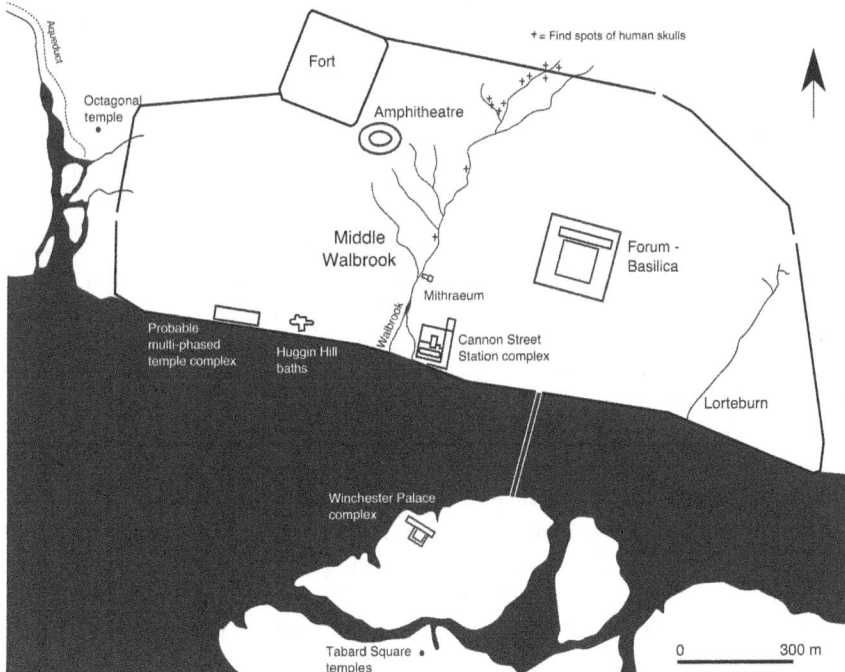

Figure 3.8 Roman London with the defining spatial influence of the Thames and Walbrook.

Source: Drawn by author after Rowsome (2008) and Rogers (2011a).

shows evidence of cult worship of other exotic deities such as Sabazios. The worship of both Mithras and Sabazios often involved water, with snake imagery linking to more traditional deities like Aesculapius (ibid.: 120). Furthermore, a great deal of metalwork has been recovered from the Middle Walbrook area, with a strong preponderance of dress/ personal items deposited throughout the first and second centuries (Crummy and Pohl, 2008: 212). Merrifield and Hall (2008) see this as evidence for a continuation of the aforementioned votive tradition.

It is important to note that this area required extensive and continuous land reclamation during the Roman period, meaning that even the *mithraeum* would have been built on rather insecure ground (Rogers, 2011a: 212). In addition, the Middle Walbrook represents the tidal head of the watercourse, so it was an area where there was visible difference in water flow. Progressive reclamation of the area could have led to braiding of the river channel and the creation of island-type areas; it is possible that the *mithraeum* itself was bounded either side by water. Intriguingly, prior to establishment of the temple to Mithras this zone of activity is framed by the creation of two Roman bridges over the Walbrook. Merrifield and Hall (2008) connect these structures with the ongoing ritual activity of the area, with a collection of small metalwork item deposits in the river at these crossing points.

This interpretation is reinforced by the routes of the two roads after crossing the Walbrook. Their passage over the River Fleet, in the western area of Roman London, coincides with notable junctures of that watercourse. The northerly road crossed the river at the point where the channel widens, which again seems to be a tidal head; the more southerly road crossed between two eyots that had formed in the mouth of the Fleet. These points are further places where the flow of water would have been visibly affected. Added to this, Crummy and Pohl (2008: 219) highlight a series of possible structured depositions involving toilet instruments, similar to those found in the Middle Walbrook, close to a jetty on the southern eyot of the Fleet. Such small items are known to have been used as *ex-votos*, and their good condition suggests they were not discarded as rubbish (ibid.: 218). There is also a later construction of a large octagonal structure, thought to be a temple, on a hilltop overlooking this area of the river (ibid.: 219). It seems possible to consider the bridges as structures that enhance the meaning of an area already conceived of in special terms. Indeed, as with the Thames bridge, the crossing points of the Fleet and Walbrook could have a significant role in the ideological unity of the settlement.

Water supply and the Walbrook Valley

In the Walbrook Valley there is a large concentration of wells (Figure 3.9). Three of the most impressive examples were found at Gresham

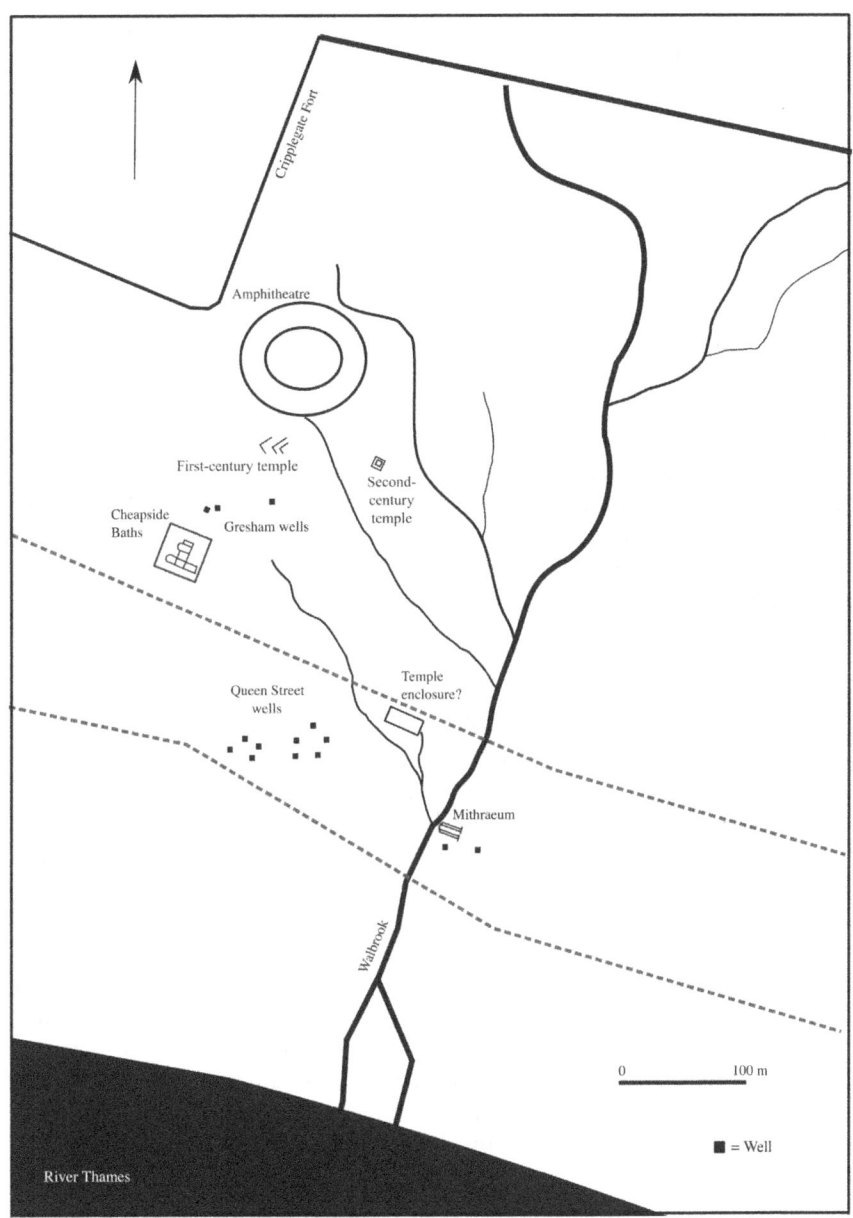

Figure 3.9 Locations of wells in the Middle Walbrook.

Source: Drawn by author after Rowsome *et al.* (2011).

Street, in the modern City of London and close to Cheapside Bathhouse (Blair and Hall, 2001; Blair, 2002; Williams, 2003). These structures provided a significant amount of water for high-demand parts of the city. The remains of sophisticated lifting equipment have been recovered from within the wells, leading to numerous theories as to how they functioned. Archaeologists have sought a better understanding of how such Roman wells were constructed, how machinery like these lifting devices was designed and how much water the wells could yield. In part this is due to the long-running debate on whether London needed an external water supply (Wacher, 1978; Wilmott, 1982a; Williams, 2003). While this is a fascinating line of thought, as these lifting devices are rarely recovered anywhere in the Empire, we cannot underplay the diverse social reaction that could have characterised the introduction of such technology.[6] As with many elements of Roman water supply, archaeologists seem to have been pulled into a familiar casting of these wells as primarily achievements of Mediterranean ingenuity and progress. The work of Oleson (1984) appears a number of times in the interpretation of London's lifting devices. His dense and ranging narrative of the history of such engineering structures in the classical world is essential to consider; yet it also undeniably moves the focus away from local diversity and provincial social reactions.

In some ways the problems of trying to understand these structures by using modern engineering have been exposed by this process. Blair (2002: 516) noted, reasonably early in the project at Gresham Street, how a massive team of engineers and mathematicians were engaged in trying to recreate the lifting devices of the wells, but acknowledged that such modern techniques would not have been available to the Roman worker. On the surface this seems to be merely an acknowledgement of Roman achievement, but it underlines a crucial realisation: when creating these wells, Roman builders were not merely thinking about equations and data, but were drawing on their traditional engineering/craftwork techniques, their past personal experience, their knowledge of specific materials (which may have had 'special' properties) and, of course, on the blessing of their gods. The last factor is surely crucial when we think about the explanation of failure. Indeed, while we may interpret the ultimate failure of these wells as resulting from the 'trial-and-error' nature of their construction, it was not necessarily the view taken by the populace of London at the time – they may have been more likely to attribute this failure to divine intervention. This is somewhat corroborated by the fact that two of these wells appear to have been in use at the same time; there is not a sense that one was a definitive second version of the other.

In this vein, Gerard (2011) analysed the findings from two late Roman wells (albeit not on the same scale as those found at Gresham Street) at the Draper's Garden site in the Upper Walbrook area. At the bottom of these features was a remarkable collection of copper alloy, lead alloy and iron vessels, as well as an interesting faunal assemblage (ibid.: 552). This rich haul of items led archaeologists to interpret the wells as being part of marked ritual activity. However, Gerard (ibid.: 563) is keen to point out that finds such as an iron bucket binding may appear to be entirely utilitarian (and thus tell us less about beliefs), but should not be underestimated. He notes how these elements could have possessed significance because of their long association with the function of the water source. It is possible that the length of time these items were part of the functioning well may have imbued them with importance beyond that of the more immediately impressive items found at Draper's Gardens. Such an interpretation affects our conception of the monumental wells in the Middle Walbrook. The advanced lifting equipment recovered, which has been the focus of intense engineering and scientific postulation, could have been invested with just as much meaning as traditional 'votive offerings'. Blair (2002: 513) makes a point of emphasising the craft involved in producing the water cups, the part of the well mechanism most intimately related to the water itself. It seems plausible to suggest that such items may have accrued a status beyond the mundane confines of practicality.

Analysis of the two larger well shafts at Gresham Street has shown that both were the victims of rather calamitous ends. The first seemingly collapsed in on itself due to structural weakness; the second was probably destroyed in the Hadrianic fire of London (ibid.: 512). Considering that the items we describe as votive offerings are often found in wells as a product of a structured backfill (perhaps involving a ceremony of sorts), we do not necessarily have the right circumstances for comparison. Furthermore, the unfortunate fates that befell the structures have to be viewed through the perspective of a superstitious audience. The events leading to the destruction of the wells could have been perceived as proof of supernatural intervention, in a similar vein to the ideas discussed in relation to the probably non-functional well in Building 8 at Verulamium. It is plausible that this could have reaffirmed a sense of difference for these examples (in contrast to other wells in London). Such unprecedented interaction with the underground water of the area could have been perceived as an intrusion on a scale that was unacceptable for the gods. The fact that there is no attempt at repairs, and one well is not a direct replacement of the other, also fits this theory. Furthermore, the third well in this area is known to be later, but it is notably smaller. This line of thinking opens the way for interpreting this as a return to a more familiar scale of water interaction.

A further group of wells have been found in an area around modern Queen Street that seems to have been continuously used for this purpose (Figure 3.9) (Wacher, 1978: 107). The excavation at this site alone revealed 18 wells, 13 of which were from the first and second century

(Wilmott, 1982a: 240). In geological terms it was a zone located on the terrace of the westernmost of the two hills on which the city was built. Wilmott (1982b: 5) notes that the fact of the London clay being at a higher level and the overlying gravels being less thick created an ideal situation for access to the underground water. Combine this with the apparent lack of votive deposits, and it is not hard to see why this area has been perceived in strictly practical terms. Wacher (1978: 107) asserts that the water could have been used for public provision. The nearby site at Watling Court appears to have been a densely populated residential area, which would have required a continuous water supply. The main point emphasised in Wilmott's report is the idea that this area could have been pivotal in the overall water supply system of Roman London (Wilmott, 1982b: 16). Nevertheless, concentrating on these issues has perhaps obscured the broader significance of having an area at Queen Street that is devoted to water.

One well in this area appears to have a traditional votive deposit: a human skull was found in it (ibid.: 9), which obviously recalls prehistoric-type deposition in watery contexts (not least the Walbrook). It is interesting that this particular well was cut through the gravel cap into the London clay, an exceptional example among the wells of Queen Street. On a practical level this meant water from the gravel could flood into the well, providing a substantial amount of clear water at all times (ibid.: 5). However, this well is probably the closest of these structures to the Walbrook, perhaps linking it with the similar traditions of skull deposition in that watercourse during prehistory (see Bradley and Gordon, 1988; Lees, 1989). The profile of the well shows a deeper intrusion into the London clay, which could be seen as a more profound interaction with the 'other world' found beneath the ground. Furthermore, in a very stoical sense the extra effort required to delve deeper could have resulted in more meaning for the structure in its working lifetime.

We are quite familiar with the idea of wells being located in small groups near other buildings; these are often defined by their surroundings, or what we find in them. However, the case of Queen Street provides us with an example of wells genuinely defining an area of a major Roman town. Indeed, the inference here is that people needed to frequent this area and interacted with the wells on a consistent basis. The association of this place with water was reinforced through sensory perception: the sound of these wells, the movement between them and the various flavours of the water (perhaps some were more popular in this regard) would all have played upon the individual. This may have been a profound experience that linked the individual with the immediate waterscape, already illustrated as being ritually charged. If we think about the celebration of the Fontinalia festival, mentioned previously, the impact of the Queen Street area could have been significant. The concentration on well dressing and spring veneration would have transformed this area of the city. At the very least, it could have been one of the most influential points in the urban landscape on one occasion each year.

A London aqueduct?

While wells appear to have supplied the vast majority of water to the settlement, the possibility of an aqueduct water supply for London has been a source of debate for archaeologists. In light of the recent discoveries of monumental wells (Blair, 2002), there is increasingly a consensus that the city did not possess a significant aqueduct from an external source (Blair and Hall, 2001; Williams, 2003), although the dense modern occupation of the capital makes it hard to form firm conclusions on the matter. While there may not be significant evidence for an aqueduct leading directly into the central areas, a leat-style conduit has been discovered on the outskirts, in association with the River Fleet. It siphoned water from further upstream to the point of the small eyots at the Fleet crossing (see Figure 3.8) (Spain, 2004: 41–42). This conduit has been viewed as being part of an industrial zone, and the power of the water flow was directed into water mills (ibid.: 47). While this area may have been involved in manufacture, it also seems to have had a ritual meaning.

The issue of water mills is a topic that needs some discussion and clarification. They invariably require the construction of a channel to focus the power of a nearby watercourse, so they are frequently linked to the creation of aqueducts: the Barbegal mill close to Arles in France is a case in point (Leveau, 1996). The problem is that because mills serve a seemingly clear purpose of production, their presence can often be seen as a force to count against more ritual or ideological interpretations. Yet Purcell (1996) notes how the idea of a 'productive landscape' in Italy often had more to do with symbolic processes of transformation, hence productive villa landscapes were not always located in places where maximum agricultural yield could be attained. Sometimes distinctly unhelpful, often waterlogged, locales were selected and then transformed. This was agricultural production, but set against the background of an unfavourable setting. In this way, the ideological potency of the villa was greatly enhanced.

Thus a leat system, although patently possessing a practical function, could also have other associations based on the significance of the area and the medium of water. Even on a surface level, one can certainly suggest this structure may have changed the flow pattern of the Fleet, and the potential significance of this kind of change has already been outlined. However, with the two bridge crossings into the town, changing tidal points, islands and the conduit, we have a location that is defined by watery transformation; the natural elements intertwined with the human-made features, enhancing each other.

With this in mind, it is worth considering the Walbrook, with its legacy of prehistoric meaning. The extent to which it became something resembling a leat in the Roman period is particularly intriguing. Various accounts point towards the Middle Walbrook becoming an increasingly canalised and human-made structure. At modern Bucklersbury House evidence has been found of continual Roman projects to maintain the banks of the Middle Walbrook: large timber revetments constructed towards the end of the first century, and the artificial raising of both banks via a coordinated programme of clay and gravel dumping. Rich organic material found on the banks has been interpreted as the upcast from cleaning of the channel itself (Hill and Rowsome, 2011; Rogers, 2013: 51). This sort of activity is a small-scale example of the river hybridisation that Edgeworth (2011: 77) identifies in watercourses like the Yellow River in China, but it also emphasises the problems inherent in seeing aqueducts as entirely detached from rivers during the Roman period – the Walbrook was even being cleaned in a similar way to aqueducts. One has to wonder whether such activity was linked to different festivals and celebrations.

The Middle Walbrook area has also yielded much evidence for water piping, which suggests a local network of supply. The stream features so heavily in the topography of the valley floor that it is plausible to think such a network would incorporate the Walbrook as a primary source. At times there was industrial activity in the area. The evidence of metalworking is perhaps not as apparent as one might expect, but there are definite pointers towards milling being undertaken: a well-built timber cistern was found close to the Walbrook in association with an unusual amount of quern stones (Hill and Rowsome, 2011: 412). Possible piping of this water back towards the main stream could have created enough force to drive mills. The sheer amount of activity involving water in the Walbrook Valley makes it an intriguing centrepiece to the Roman town. Surely the effect of weaving this highly significant watercourse so deeply into the urban experience would have a powerful ideological effect.

While it seems unlikely that London ever received an external aqueduct, the internal network of water piping was apparently extensive. In addition to the piping in the Walbrook area, examples have been found in Southwark, the forum area and further west towards the Lorteburn (Milne, 1992: 23; Hill and Rowsome, 2011: 413). As already emphasised, even before Roman occupation this area was a fusion point for numerous different watercourses, so the prospect of a complex network of water pipes adding to this situation is notable. Furthermore, this water would have been channelled into the very fabric of nearby buildings, meaning it may have had a direct effect on how people experienced the structure.

Bathing in London

From a traditional point of view the marked presence of water in Roman London, coupled with the wealth of the settlement, makes it unsurprising that there were a number of bathing facilities constructed in this period (Figure 3.10). However, given the ritual and symbolic potential of water outlined above, we should pause for thought before attributing a generalised meaning that does not consider the local context. With this in mind, it is noteworthy that the zones discussed above (Southwark, the Thames waterfront and the Walbrook Valley) correlate strongly with the location of important bathhouses. While there are practical reasons for that distribution, it is somewhat surprising that the huge investment in London's forum and basilica site did not result in the creation of a monumental centralised bathing facility in that area. There is evidence in at least 12 separate locations in London that has been interpreted as indicating public or private bathing facilities, but this figure continues to rise with development-led archaeology in the city.

The Thames waterfront was a clear focal point for these structures, and the Huggin Hill bathhouse would have been a major presence. This

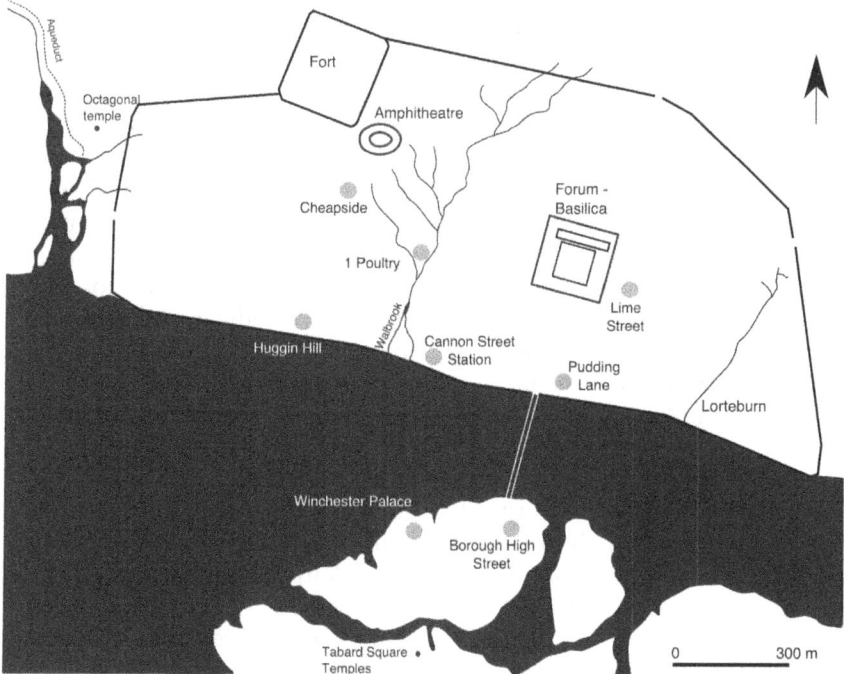

Figure 3.10 Locations of bathhouses in Roman London.

Source: Drawn by author after Rowsome (2008) and Rogers (2011a).

building had a typical public bathhouse format, and was probably one of the first prominent monumental structures of the town. It was built on a series of terraces painstakingly cut into a steep hillside close to the Thames. At this point the geology was particularly vibrant, with an active spring line existing at the interface of the gravel and the impervious clay (Rowsome, 1999: 263). This meant the bathhouse had a constant supply of water – probably incorporated into the structure with cisterns and piping (ibid.: 264), as seen elsewhere in the Empire. However, it also means that the building was situated at a transient point in the landscape; the efforts required to build in this location imply that the result was worth the initial toil. The building was monumental in scale and would have dominated the riverfront between the Walbrook and the Fleet. At high water the Thames would have reached the back of the structure; where the discovery of massive foundations could be evidence of a large landing stage. This presents us with the possibility of entering the establishment from the river (ibid.), thus for some individuals it was potentially a distinct point of arrival at the early town.

The lifespan of this bathhouse is somewhat hard to discern. Established in the first century, Huggin Hill could have fallen out of use as a primary bathhouse by the mid-second century or before. This is often seen as an index of the decline of the early vibrancy of London, and suggestive of a decline in population from the later second century. Yet there were noteworthy additions to the structure throughout its lifespan. This has been discussed as evidence of a strong public demand for the services of the bathhouse and subsequent need for its expansion. The addition of a new *caldarium* in the Flavian era represented the most substantial works; the necessary realignment of heating and drainage would have been particularly demanding. But despite the investment throughout the first century, by the Hadrianic era at least part of the structure was being used for other activities. In some of the minor rooms internal buildings had been constructed, with evidence of iron working (ibid.: 270). Nevertheless, the main vaulted sections of the structure could well have functioned throughout the period. Indeed, Rogers (2008: 141) notes how some of the structure's walls remained extant into the medieval era of London: ninth-century records indicate a building called the Hwaetmundes stan in the area of the bathhouse (Dyson, 1978: 209). Instead of decline, Huggin Hill might be tentative evidence of the hybrid nature of baths in Britain. Potential symbolic and ritual links to the water of the area, in addition to other activities like metalworking, could all have been seen in this space.

At least three other bathhouses have been found close to the riverfront, including the remains of a building in Pudding Lane which has been interpreted as being a second-century bathhouse (Perring, 1991: 73). Its location is again below the spring line, close to the Thames foreshore; it also happens to be at the bridgehead of the primary (Roman) Thames bridge.

Once more, this represents a space for social mixing on a primary thoroughfare to the forum, and a focal point for water close to a bridge that has already been discussed in ritual terms. While the practical attributes of a bathhouse in this location are evident, it is worth considering whether we can detach water veneration on the bridge from a nearby building so intimately related to the same medium. It is made more interesting by the 2011 discovery of a likely bathhouse in Southwark at 11–15 Borough High Street, directly on the other side of the old Roman bridge (Fairman, 2013). While such facilities could be interpreted as establishments built to cater for the many travellers coming into London via the Thames bridge, the presence of ritual activity in addition to ideas about procession on the bridge towards the forum opens the possibility for a variety of purposes. Fairman (ibid.: 150) notes that the Southwark bathhouse appears to have been subject to numerous structural alterations, suggesting multiple functions. Moreover, its situation in the upper limits of the northern eyot of Southwark was noted as probably deliberate positioning to promote visibility in the landscape (ibid.). The same may be true for the bathing facilities at the Winchester Palace site in Southwark (Yule, 2005). Their location on the northern island overlooking the main flow of the Thames, and possibly in view of the Huggin Hill baths, may have played an essential role in the experience of the building in the Roman period.

The Walbrook Valley is the other place where baths appear to have proliferated, and in many ways the facilities here may have been more successful and fundamentally important to the inhabitants of the city. The first place to mention is the Cannon Street complex, which was an ambiguous monumental structure at the mouth of the Walbrook; the archaeological evidence relates to numerous components of different dates which have been interpreted in various ways (Milne, 1996). This structure was traditionally labelled as the Governor's Palace (Marsden, 1975), but increased understanding of the area precludes this as a possibility and the nature of the reveted Walbrook in the Roman period means an originally assumed western wing was unlikely. Regardless, the central part of the complex included a large hall and an ornamental pool, both of which occupied a podium of over 100 square metres (Rowsome, 1999: 274). While this does not exactly provide profound evidence of a public bathhouse in the traditional sense (there are no rooms designated explicitly as *caldarium*, *tepidarium*, etc.), the pool is most fitting in such a structure. Moreover, there have been discoveries nearby (to the east and west) of substantial stone buildings with hypocausts. The evidence recovered at 3–7 Dowgate (just northeast of the mouth of the Walbrook) is particularly reminiscent of a small bathhouse (ibid.). What this seems to imply is that this area, of great importance in the wider waterscape, had a series of buildings with water as their focal point. In the case of the Cannon Street complex this was a monumental construction dedicated to water at a place where the prehistoric meaning of water was evidently particularly acute.

A first-century bathhouse was located at Cheapside, close to the Cripplegate fort and later amphitheatre (Perring, 1991: 73). This structure is touched upon in the discussion of the monumental wells at Gresham Street, which could have served as a viable water source. The baths seem to have been in use up to the early third century (Rowsome, 1999: 274). While it has often been said that the proximity to London's fort would have meant a predominantly military clientele, it is worth reaffirming the importance of the surrounding area. The excavations relating to the Gresham wells uncovered a bronze arm from a large statue (Bayley et al., 2009). This was interpreted as a bronze of Nero or another early emperor that could have stood in front of a nearby temple (Rowsome et al., 2011). This location on one of the principal coaxial roads of London befits such monumental structures, and therefore might suggest a more opulent public balneum.

Another small bathhouse has been interpreted at the 1 Poultry site in the City of London (Hill and Rowsome, 2011: 370). There is some debate as to whether this was a private residence or a small commercial bath building. As a structure its use endured well into the fourth century, implying a degree of success (Rowsome, 1999: 275). The arguments against the bathhouse interpretation essentially centre on the lack of a readily detectable room sequence in alignment with classic bathing practice (Hill and Rowsome, 2011: 371). Of course, as mentioned earlier it is debatable whether compliance to such orthodoxy is a fair assessment of a bathing facility. But if this building was a bathhouse it would have been another prominent water feature in this area close to the Walbrook, in addition to the bridge and the dedicated well zone already noted. The same aspects mentioned for Pudding Lane could be applied here concerning the potential ritual activity attributed to the nearby Walbrook.[7] If this water harboured meaning from any Iron Age (or earlier) associations, the combination of these features needs further scrutiny. At the very least, the emerging affinity for these structures could be a legacy of the similar social aspect of water in preceding eras.

In this regard it is worth noting that some of the small, potentially votive, offerings found in the Walbrook were directly related to bathing. Wardle (2008: 209) noted that at least 12 iron strigils have been found on sites in and around the Walbrook, raising the question of whether they could have been part of ritual behaviour. These implements are all in full working order, and conspicuous in that they represent almost the complete collection of such items found in London (ibid.: 202); the only other strigil in the British Museum's permanent collection was found in a waterlogged deposit in Southwark. While the conducive preservation conditions of the Walbrook Valley temper interpretation somewhat, we cannot dismiss their potential as evidence of ritualistic behaviour taking place in the area.

The other interesting aspect is that strigils are usually made out of copper alloy (Manning, 1985). Iron deposition has a strong tradition in temperate Europe, and thus this selection of material could have been a deliberate decision.

Many of the towns explored in this book were not only sited close to interesting points in their local waterscape but also actively engaged with them. London is a prime example of urban water features highlighting these places of symbolic potential in the local waterscape and integrating them with the fundamental experience of the town. The establishment of numerous bathhouses seems to solidify this close relationship between important points of the waterscape and the urban experience. The Huggin Hill baths were located on the spring line close to the Thames and were part of the panorama of the symbolically charged riverfront – a locale that increasingly gained importance throughout the Roman period (Rogers, 2013). Other baths have been found close to the Thames bridge, on the northern Southwark island, at the confluence of the Walbrook and the Thames, and in the Middle Walbrook zone. These examples thus appear in the areas where water may have been at its most meaningful and symbolically powerful. When we also acknowledge the ritual connections of their water supply, perhaps coming from wells with votive deposits or the culturally charged watercourse of the Walbrook, it becomes difficult to describe their function in generalised terms.

Silchester (Calleva Atrebatum)

An underground waterscape

In contrast to many of the towns mentioned in this chapter, Silchester was not sited close to a prominent river. However, there were probably a number of small brooks in the area, and indisputably a plentiful supply of underground water at the site (Boon, 1974: 85). Without an obvious river source, the procurement of this groundwater would have been an essential part of life in the Iron Age and Roman period. So it is of little surprise that this Roman settlement has produced evidence of more wells than any other site in the province, although this fact is tempered by the knowledge that the site is also one of the most extensively explored given its 'greenfield' nature and thereby accessibility to the archaeological spade. While practical requirements are apparent, there was undoubtedly a ritual element to some of these structures.

The antiquarian excavations in Insula IV (the area of the forum; Figure 3.11), primarily by Fox and Hope (1894), uncovered at least four wells, all with probable votive items. Among the finds were a bronze handle, two perfect pots, a lead weight, a steelyard weight, flints, coins, pewter cups, an iron stylus, two jawbones of cattle and dog bones. These finds in wells

Figure 3.11 Roman Silchester with notable features.
Source: Drawn by author after Creighton (2006).

were accompanied by many similar deposits in pits, such as the skulls of four dogs, spurs of gamecocks and a small knife blade all found buried beneath the forum (Joyce, 1881). Fulford (2001) mentions five other insulae that contain at least one well with unusual deposits.[8] One of the most striking of these was found in Insula XXIII; it held a hoard of more than 100 pieces of ironwork, as well as bronze vessels and two complete pottery vessels (Fox and Hope, 1901: 246–250). These items were recovered above a 1.5-metre-thick deposit of black ash, which appears to suggest some dramatic event or ritual burning. In some ways this deposit is less useful because it was from a later period (fourth century), but in conjunction with the other evidence it may imply enduring traditions associated with wells in Silchester.

These unusual well deposits occur in almost all the central insulae of the town – the areas we associate as most overtly Roman and institutionalised in the settlement and surrounding *civitas*. They underpin the symbolic heart of the site in a very literal sense. The modern excavations led by Fulford have further underlined their importance in this area from the Late Iron Age onwards (see Fulford and Timby, 2000). Seven features found under the Roman forum-basilica were deemed to be wells; their initial construction was dated to the period between 25 BC and 15 BC (ibid.: 8), meaning they were presumably central features of the pre-Roman settlement at Silchester. Of these seven wells, four were still in use around AD 50 and two continued into the Claudio-Neronian era. Another (Well F192) was probably dug around AD 50 and remained for a much longer time, deep into the High Roman period.

Some of the most striking objects found in pits and wells in Silchester have certainly come from later periods (see Eckhardt, 2006), but there is evidence of notable ritual closing of wells from the Late Iron Age. Well F423, for example, in the central area of the town, contained an exceptionally rich ceramic assemblage, with a variety of Central Gaulish and Gallo-Belgic imports stamped Arretine of Augustan-Tiberian date, and an uninscribed Trinovantian copper alloy coin. Furthermore, other finds included a bone spindle-whorl, two perforated pottery discs, a small clay thumb-pot, a fragment of a glass bowl, an iron awl, a nail and an iron brooch (Fulford and Timby, 2000: 17).

In more recent excavations of Insula IX, a number of wells with a long chronology have been linked to ritual activity. Well 10421 had a primary fill dating from circa 20–10 BC, and a complete courseware cooking pot and a partially complete Silchesterware beaker were placed in it before backfilling (Lodwick, 2015: 101). Similarly, Well 8328 may have been sunk to replace 10421 in the early first century, and was filled with fragments of quern stones and a complete pierced pot, alongside a deposit of stable manure. Within the fill there was evidence of plants that have been connected to rituals – including *Hyoscyamus niger*, *Malva sylvestrism* and *Solanum nigrum* (ibid.: 102). Fifteen metres to the south, Well 1586 was sunk in the post-conquest period and also has evidence of several intact pots and flagons (ibid.: 103). This suggests a set of practices connected to water supply that, while perhaps becoming more elaborate in later periods, certainly had a long history. Certain aspects may have continued relatively unchanged. Lodwick (ibid.: 107) notes that the inclusion of stable-manure assemblages of plant remains in the fill of wells can be seen in early and late features. Such a deposits had a strong sensory connection, and may have been just as profound as human-made objects.

As highlighted by Fulford and Timby (2001), there is also a 'pervasive' pattern of deposition involving 'holed' pots; upwards of 70 examples of such vessels have been recorded in the Silchester Collection. The vast majority of these appear to have been placed in wells or pits with holes in the belly of the vessels (ibid.: 293). Jars make up nearly half of this number, with flasks, beakers and bowls also prominent. Some of these are substantially complete, despite being at the bottom of a deep well, suggesting careful placement. The reason for this activity is uncertain, but it could represent a type of ritual 'killing' of the vessel (multiple holes in the belly/base making it useless for carrying water). However, the practice could also be evidence of other activities carried out in association with wells. Fulford and Timby (ibid.) suggest a possibility that the containers could be timing devices. It is plausible to suggest such vessels were used to test for well purity; a carefully holed jar could release water but retain unwelcome sediment for inspection. Moreover, the sound of this water being released could give a rough estimate of depth to ascertain whether more accurate measuring was required.

As mentioned previously, awareness of such attributes could be a gauge of how such a feature was perceived in the local belief system. There is substantial evidence for this trend, spanning from the Late Iron Age into the fifth century AD (ibid.: 294). Of course, this appears to imply a degree of continuity from the Iron Age in the practice of interacting with wells. Considering the other notable deposits in these features, it is certainly possible that this piercing of vessels was part of an expression of ritualistic action. The fact that many of the wells were present in spaces that became central 'Roman' features (e.g. the forum-basilica) means one must question how people in Silchester were interpreting such spaces.

Water and monumental buildings in Silchester

In terms of a piped water supply that we generally associate with provision of water to monumental structures, the bulk of the evidence in Silchester is centred on a wooden pipeline associated with the southwestern town gate (Hope, 1897: 422). A ditch was discovered with iron collars leading from the town wall through Insulae XV and XVI and into III. At the wall, evidence of the conduit ended at a rough mass of flintwork, which could have been the foundation for a water-tower to raise the supply (Boon, 1974: 88); such a structure would presumably have been required, as the only close source of water was at a low level compared to the town itself. This could have been an early well, possibly supplying a small bathhouse in Insula III. While the evidence is not especially comprehensive, it nonetheless remains an intriguing feature, especially bearing in mind the copious supply of well water in the town. Indeed, some of those features were subsequently also the source of the piped water supply – they created a sort of internal aqueduct system. For instance, an unusually large and well-made oak trough was traced running southwards from a well in Insula VI. This could feasibly have linked with other similar timber-trough structures that have been discovered in the nearby insulae (ibid.). The discovery of a force pump is also indicative of advanced lifting of spring water to provide for structures such as the bathhouses (ibid.).

The primary public bathhouse at Silchester was probably one of the earliest examples of such a structure in the province, being built in the first century, and bearing some similarities to both London's Huggin Hill baths and the Jewry Wall baths at Leicester (Burgers, 2001). The building likely started out as a simple row-type bathhouse before extensions and additions made it a more complex structure. There is some mark of local social change in the improvements made to the layout. During the second century a new *caldarium* was constructed, hinting at the possibility of both men and women bathing at the same time or that the baths began to attract a much larger clientele (Boon, 1974: 130). Yet the most intriguing element of the bathhouse might be the build-up of other aspects of water significance in its immediate locale. Insula XXXII contains the primary bathhouse, a stream possibly supplying it, at least one well with a votive deposit and a suspected *mithraeum* (of a later date) (ibid.: 156). While not much has been made of this watercourse, it would have been a feature of the former Iron Age *oppidum* and thus would have local associations. Indeed, the lack of a prominent river running through (or by) the town lends credence to the notion of its enduring importance.

The ongoing excavations at Silchester have revealed that the Late Iron Age occupation of the site seemed to influence many of the early monumental buildings of the Roman era (Creighton, 2006: 65). One of the most marked examples of this is the public bathhouse (Fox, 1948), which was later altered to fit in with the slightly different alignment of the Roman street grid. This is an intriguing point, because it frames the creation of the bathhouse within the pre-Roman layout and conception of space. Creighton (2006: 141) observes that the structure is on the same alignment as the nearby temple complex, and both were outside the Iron Age earthworks that defined the settlement area. Unfortunately, our knowledge of this area has been hindered by the presence of modern buildings; yet potentially the *temenos* and bathhouse structures could represent a similar set of circumstances to those observed at Verulamium and Folly Lane (ibid.). While Creighton emphasises the impact of a similar external temple and bathhouse connection at Silchester, there is the added parallel of water being a conspicuous presence in the immediate area. The poor understanding of the *temenos* enclosure may even disguise further interactions with groundwater. There are at least two Romano-Celtic style temples within the enclosure that are of an early date (ibid.), and the influence of local custom on their function must have been marked.

The other notable bathing facility of the town is located within the supposed *mansio* (Insula VIII). Wacher (1995: 278) suggests that while this large construction could have been part of a personal estate, it was more likely intended to cater for the needs of the high volume of travellers passing through this prosperous trading town. While the provision of a *mansio* with bathing facilities is not unusual, the size of these is somewhat at odds with other British evidence. The similarly early date for these baths suggests that they were running at the same time as the public baths. As such, maybe the public baths had a more pronounced connection to local customs, while the *mansio* baths accommodated more practical needs of the traders and the transient population.

In addition to the forum and baths being closely aligned to water sources, the amphitheatre, located just outside the eastern town walls (but inside the surrounding earthworks), was closely related to a nearby spring. A large coping stone found beside this water source suggests the possibility of an ornamental pool or surrounding structure (Boon, 1974: 159). A strong link between water and amphitheatres is not unheard of in the territory surrounding Silchester. At Frilford, for example, a structure was built in a boggy landscape with a possible spring incorporated (Rogers, 2011a: 98). Indeed, a feature originally interpreted as a 'royal box' could, in fact, have been part of a construction that directed the flow of this spring within the arena. Frilford has been seen as a rural religious centre, with a prominent *temenos* that was probably constructed due to the watery nature of the site (Hingley, 1982). Use of the amphitheatre appears to have been related to this ritual/religious function, rather than the more traditional entertainment often ascribed to such buildings. This relationship to the landscape was perhaps further ingrained by the positioning of the amphitheatre at the bottom of a shallow dry valley which runs south towards the River Ock (ibid.). While nothing as convincing pertaining to ritual function has been found in association with Silchester's amphitheatre, the overall value of underground water in this settlement should make us question the proximity between such structures and water sources.

Silchester offers us a slightly different type of waterscape. The sheer number of wells with remarkable deposits gives clear evidence of a special connection between the people who lived here and their underground water. To an extent we have limited interpretation of these wells just to analysis of the deposited objects, rather than of the water itself. These votives are a symptom of associations with the water, which was being moved out of these wells and into the fabric of the monumental buildings of the town – whether by the hand of individuals or the more sophisticated pumping equipment they constructed. The appearance of striking water deposits should make us question how certain buildings were experienced. The bathhouse, with its already conspicuous alignment with the nearby *temenos*, is a principal example. While we have been content to see this as a British *thermae*, showing us the changing tastes of British people, it could feasibly have been more akin to a new development in the religious connections of this immediate area – a new temple to water for a local people clearly concerned with ritual acknowledgement of their underground waterscape.

Dorchester (Durnovaria)

Water and prehistoric Dorchester

Archaeological evidence of occupation in the area of Dorchester is clear in prehistory, even if the actual site of the Roman town has fewer markers than sites mentioned earlier, such as St Albans or Silchester. The remains of the Iron Age Maiden Castle, to the south, are striking. There was also Poundbury hillfort to the east, which was likely constructed in the Bronze Age and occupied into the Late Iron Age. These hilltop settlements, particularly the former, have dominated discussions about the nature of Roman conquest in the region. The notable 1930s' excavation of the site by Tessa and Mortimer Wheeler, while revealing evocative archaeological evidence of Iron Age life in the area, has undoubtedly skewed interpretations towards issues of conflict with the incoming military power of Rome. Of the 52 bodies exhumed at Maiden Castle, 14 showed signs of trauma (Papworth, 2011: 77) and 1 had an iron projectile embedded in its back – an object that was interpreted as a Roman ballista (Wheeler, 1943: Plate XVIII). Since the Durotriges tribe of the region is mentioned in the classical historical narrative, such evidence was quickly linked to this traditional story of native resistance. Accordingly, the abandonment of Maiden Castle and the establishment of Roman Dorchester were seen as principally motivated by incoming Roman goals rather than being part of the ongoing local narrative of the region.

However, more recent work led by Niall Sharples (Sharples and Ambers, 1991) showed that Maiden Castle was probably already in decline as a centre of population by 100 BC, suggesting the local populace was the stimulus of changing settlement patterns. The ongoing Durotriges Project has since confirmed this notion, with the discovery of a large-scale unenclosed site of the Late Iron Age at Winterbourne Kingston (approximately 18 km northeast of Dorchester), dubbed Duropolis (Russell *et al.*, 2017). This settlement appears to have become increasingly populous at a time when Maiden Castle was declining, and bears little resemblance to the fortified central places so key to traditional conquest narratives.

This discovery, along with increased understanding of the Iron Age Dorset region as a whole, necessitates a change in the way we approach the interpretation of Roman Dorchester. The theory of a 'clean break' from the prehistoric narrative of place, driven purely by the incoming Roman power and illustrated in the shift in settlement from the markedly different Maiden Castle to a Roman town, can no longer be so easily put forward. As with other sites discussed in this chapter, discovering the truth of why the settlement at Dorchester became successful in the Roman period entails acknowledging a great deal of social and cultural entanglement. The relevance of the site for this book is that manipulation and association with water appear to play a prominent role in how people experienced the Roman town and the reason for its positioning.

The Frome, Poundbury and the Roman urban waterscape

The River Frome is an undeniably strong presence in this landscape, and its influence may be preserved in the Iron Age tribal name of the Durotriges, which has sometimes been interpreted as meaning 'water dwellers' (Strang, 2004). The importance of the river is also evidenced by the large number of assumed votive offerings recovered during the nineteenth-century lowering of the riverbed (King and Woodward, 2003: 152). There is some debate as to when the Roman settlement was established, but evidence of large-scale water structures, such as the aqueduct, a bathhouse and associated drains, indicates a prominent town close to the Frome by the late first century. As outlined in previous chapters, the creation of such features is often held up as evidence of an incoming Roman presence which had little relationship to the prior engagement with the local landscape. However, the examples we find at Dorchester are remarkable in their physical connection to prehistoric features.

The Roman aqueduct was a rather modest leat structure that tapped waters upriver on the Frome from a dam at Littlewood Farm (Burgers, 2001) (Figure 3.12). Stephens (1985a: 203) labelled this 'the least satisfactory sort of aqueduct tapping the least satisfactory of sources'. This comment is typical of the treatment of British aqueducts, concentrating on a narrow definition of function and ignoring local beliefs. Accordingly, despite extensive work on clarifying the course of the aqueduct (see

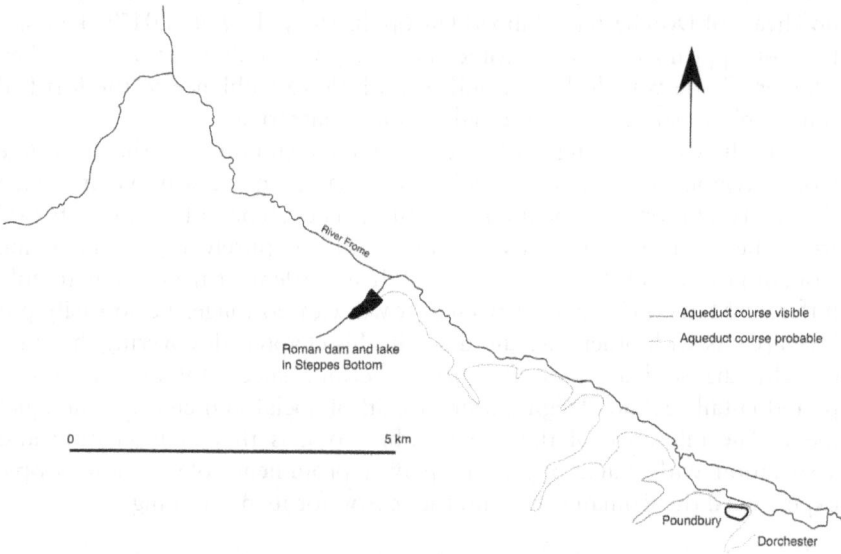

Figure 3.12 The course of the Roman aqueduct at Dorchester.

Source: Drawn by author after Putnam (1997).

Putnam, 1997), archaeologists have largely ignored the potential that it could be actively engaging with significant places in the urban periphery. Perhaps this is understandable, as for the most part the aqueduct appears to have followed a route of least resistance. However, the way that it interacts with the Poundbury hillfort is most interesting.

The prehistoric settlement at Poundbury undoubtedly had a close relationship with water, not least because it overlooks a critical fording point on the River Frome that coincides with many of the aforementioned votive deposits in the river. It is also increasingly clear that during the Iron Age the occupation of the site became concentrated on the ground between the river and the traditional earthwork boundaries of the hillfort. It appears to be only a small community, but its inhabitants are nonetheless present throughout the period in the Iron Age when Maiden Castle was a principal centre. In fact, the apparent refurbishment of the hillfort in the Late Iron Age, with the inner rampart converted from a box to glacis style and an outer rampart and ditch constructed (Papworth, 2011: 75), suggests importance for the site. Papworth (ibid.) indicates that the hillfort could be defensive insurance for the small community living close to the river in hard times. However, it seems just as possible that, due to its close association with water, this place had a high degree of symbolic importance and thus still demanded attention from the local community despite the prominence of Maiden Castle. The return to this site as the principal settlement of the area in the Roman period perhaps gives some credence to this interpretation, particularly when one considers the development of the urban waterscape of Roman Dorchester.

As already discussed in this chapter, the meaning of Iron Age earthworks appears to have been a great deal more complex than the defensive interpretations given prominence in many traditional academic accounts. With this in mind, it is of great interest that the Roman-period aqueduct was incorporated directly into the earthworks at Poundbury. Due to the structural style of British aqueducts, the leat construction is very similar to the earthworks already present. Moreover, considering this earthwork range was still being added to in the Late Iron Age, this is akin to another stage of local development rather than a radically new structure. The second phase of aqueduct construction expanded the conduit to twice its incoming size in the area of the hillfort (Putnam, 1997); the practical benefits of this would have been very limited, but the symbolism was potentially potent.

As an overt display of Roman culture and sophistication, the Dorchester aqueduct is not particularly impressive. However, as a more inclusive construction woven into the meaning of the landscape from the pre-Roman period it has potential to be a marker for a more nuanced interpretation of settlement in this locale. This entanglement between incoming Roman form and local belief structures has a direct parallel in the immediate

landscape of Dorchester. The Maumbury Rings, to the south of the
Roman-period settlement, were initially a Neolithic standing circle. During
the Roman era this ancient structure was converted into the town's amphitheatre. Some might interpret this as aggressive appropriation of a culturally important site, but we have seen in other places in Britain how theatres
in Roman towns appear to have played a role in creating new aspects to
pre-existing ritual behaviour. The path of the Dorchester aqueduct may
have been involved in similar practices, symbolically linking the wider
meaning-laden landscape with structures of the newly established town.

    This non-practical interpretation of the Dorchester water supply is given
further credence by central features of the town. Woodward and Woodward (2004) make direct comparisons between the evidence at Dorchester
and that found at Silchester and St Albans. Over the course of the Roman
period at least 19 shafts/pits/wells were dug in the central areas of the town
(ibid.: 72) (Figure 3.13). As with the site at Folly Lane, there is a degree of
uncertainty as to the exact definition and role of these features. Some seem

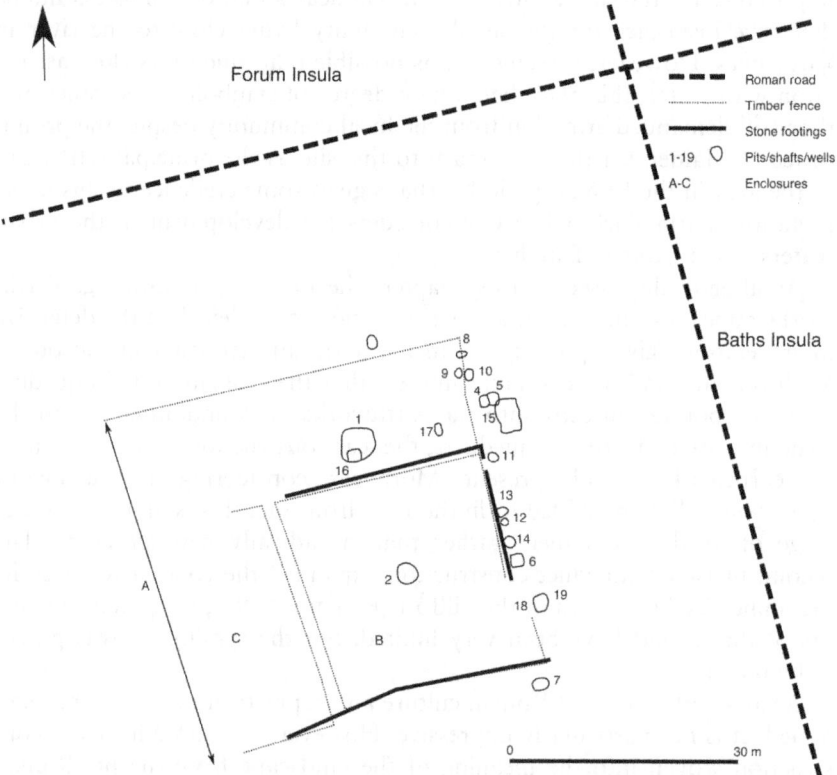

Figure 3.13 The pits in the central areas of Roman Dorchester.
Source. Drawn by author after Woodward and Woodward (2004).

more likely to be involved in water provision than others. In the second century, for example, an unusual stone-lined feature (marked 15 in the figure) probably functioned as a pool. In the fillings of the 19 pits there are numerous examples of near-complete pottery vessels, samian bowls, black burnished-ware jars/bowls and bronze jugs (ibid.). As with the previous sites, there was an exceptional faunal assemblage, including the skeletons and skulls of a large number of dogs and puppies, plus the cranium of a man in one of the deep shafts (marked 6 in the figure).

The chief interpretation of this activity has been that the pits represent urban foundation deposits within the tradition of Roman ritual practice, as engaged with by Rykwert (1976). It has also been suggested that the placement of pits/wells delineated a series of sacred enclosures; most of the pits (or wells) are clustered around the presumed edges of these areas, with the exception of two large features that were presumably focal points within the enclosures (Woodward and Woodward, 2004: 72). This type of activity is also present at the nearby Iron Age site of Duropolis. In 2015 excavations at the site revealed 18 cylindrical pits, the majority of which contained suspected ritual deposits; these included the articulated skeleton of a dog, triangular baked-clay loom weights, quern stones, upended and perforated pots and the inverted skulls of cows and horses (Russell *et al.*, 2017: 16). While this is not identical to the activity at Dorchester, it certainly seems to suggest that in central areas of the Roman town local practices are being adapted in some way. Moreover, the primary public bathhouse of the town is adjacent to these areas and has been seen as a building that was deliberately placed to engage with the pits (Woodward *et al.*, 1993).

The baths were built at the end of the first century and were of considerable size. Indeed, Wacher (1995: 325) notes how the building contained the usual selection of hot and cold rooms, but they were all exceptional in their size and number. Additionally, the structure is known to have undergone many changes, and the addition of a hot tub in the late fourth century serves to illustrate the primacy of water within its later function (Keen, 1977). It has been postulated that this feature, plus the associated shafts, is indicative of a sacred enclosure. There is a suggestion, then, that the baths were deliberately placed to associate with these powerful features within the central area of the town (in combination with the forum just to the north), creating an area with a long-lived religious focus tightly connected to the water of the area. Woodward (1993: 361) even proposes that the baths were directly sited upon a sacred spring that occupied a coombe leading down to the Frome.

The unusual layout of the baths and their longevity must be placed in this context of spatial meaning. Excavations at Wadham House and Collition Park uncovered a conduit (contemporaneous with the aqueduct and baths) heading towards the river. This feature has been interpreted as a spillway, perhaps associated with a regulatory reservoir for water that could have been situated near the West Gate. It was a massive construction by British standards, probably originally lined with masonry, for a full depth of 4 metres (Draper and Chaplin, 1982: 25). Taken together with the baths, these central structures could be interpreted as a monumentalisation of the important pre-existing landscape features and water processes. As a result, what could be seen as a relatively standard set of Roman urban developments when taken in isolation may in reality have been crucial markers of local hybrid mixing of cultural associations and building techniques.

Wroxeter (Viroconium)

Water and the Cornovii

Over the years the emphasis of archaeological inquiry at Wroxeter has been skewed heavily towards the Roman period. This is not particularly surprising considering the lack of a large-scale Iron Age settlement underlying the Roman town, although Bronze Age barrows have been noted close to the fording point of the River Severn and there appear to have been perhaps two ditched enclosures underneath the later Roman legionary fortress (see White and Barker, 1998). However, understanding the wider local pre-Roman context is complicated by the comparatively enigmatic nature of the so-called Cornovii people of the area. Not only do they have no prominent role in the classical dialogue on Britain, but they also left relatively little in the way of local pottery and metal finds – especially in comparison with some of the other Iron Age regions mentioned in this chapter (see Greene, 2015). However, while smaller finds may be comparatively hard to come by, there is indisputable evidence of large settlements in the region as a whole. The Wrekin, Old Oswestry and Caynham Camp are all impressive examples of Iron Age hillforts found within Cornovii territory. Their presence suggests a high degree of organisation and a keen interest in engaging with and remaking landscapes on a monumental scale.

The proximity of the River Severn suggests that the local populace may have had a marked connection to water, even if it was only based on the practical benefits of such a watercourse. However, there are also less apparent markers of an interest in engaging with water in this area during prehistory. The north of the Cornovii territory, in the modern county of Cheshire, is known for its brine springs; the extraction of salt from these waters has been suggested as a prominent aspect of the regional economy during prehistory and into the Roman period. Indeed, briquetage is one of the primary constituents of the ceramic evidence at some sites in the Cheshire area (Greene, 2015: 139). It is also an industry that has become synonymous with powerful and influential Iron Age sites in temperate Europe, with the salt mines at Hallstatt in Austria being the most prominent example. Its production in Cornovii territory may have been a defining aspect of the local people's culture, due to its inherent value in the preservation of food and purification/medicine, both of which were strongly ritualised within prehistoric and Roman periods. This places water as a defining aspect of this region, with the brine springs providing potential for salt production and the wider river network allowing the movement of this vital product throughout the tribal territory and beyond.

It is not just this salt industry that suggests a Cornoviian emphasis on interactions with water. Outside the hillforts mentioned above, one of the more visible traces of Iron Age land use for archaeologists working in this region are the pit alignments and other linear earthworks. As outlined at the start of this chapter, pit alignments are a particularly ambiguous monumental prehistoric feature which appears to have had a connection to water that straddles the symbolic and the physical. Rylatt and Bevan (2007) note examples in Nottinghamshire and Derbyshire where pit alignments filled with water while also often physically addressing nearby rivers. Wigley (2002: 167), in his study of prehistoric monuments of the area, notes that a similar relationship can be proposed for these features in Cornovii territory. He describes how many pit alignments in this region approach watercourses at right angles, and at their lower points such examples may have held water. Further to this, many alignments are oriented towards the heads of valleys containing springs and minor watercourses, thus highlighting these water sources and outlining potential movement towards them in the landscape (ibid.). Similar interpretations have been put forward for known dykes in the area, with many addressing watercourses, watersheds and springs (ibid.: 155; Greene, 2015: 101). While this evidence may not appear as striking as evidence we have seen at other sites, it nonetheless suggests a potential underlying logic for meaning-laden interactions with water in the surrounding landscape of Wroxeter.

The Roman town of Viroconium was founded as a legionary fortress in circa AD 60 close to a strategically critical fording point of the Severn, and thus it is tempting to suggest it has very little to do with any of the above discussions. There is no doubt that there was a sizeable military presence at the site in the early stage, but this is unlikely to be the source of continued growth for the settlement as the fortress was downgraded by the early AD 70s, and wholly decommissioned by AD 90. The transformation of the site into one of the most influential towns of the province by the middle of the second century must have had more to do with the fortress *vicus* and the role it developed for the local people (Greene, 2015: 72). This conversion from fortress to thriving town is somewhat hidden by the later archaeology, but just because Viroconium eventually developed all the trappings of a Mediterranean urban centre does not mean it is evidence of an abandonment of previous practices and beliefs.

Some results of the Wroxeter Hinterland Project may go some way to supporting such a statement. They reveal the stark contrast in Roman material culture between Viroconium and its surrounding area. Gaffney *et al.* (2007: 3) describe the surrounding rural territory as essentially 'very British' and 'Pre-Roman' in character, which contrasts sharply with towns in other areas of the country. It is an observation that appears to counter the notion that the Roman town represents the imposition of a radical new way of living for people in the Cornovii territory. On the contrary, it opens up the possibility of a local elite embracing particular cultural trappings of the Empire, but also continuing to draw on the symbolism of previous practices and organising their wider territory in a traditional way.

The form and placement of the Viroconium defences could be a bold example of this behaviour. They were made of earthen bank and ditch, but enclosed an irregular topographic interior and an area beyond the nearby Bell Brook watercourse (Greene, 2015: 81) (Figure 3.14). As a result, they did not create a defensible site. Moreover, they were never replaced by a stone construction, suggesting they were already a structure of prestige – perhaps echoing the local hillforts and their traditional earthwork boundaries (White and Barker, 1998: 98–99). The inclusion of the Bell Brook within the confines of the earthworks is a decision that one could suggest echoes some of the above activity, highlighting watercourses and their role in movement through the landscape. The Bell Brook leads down to the fording point of the River Severn, and the valley could have been a traditional pathway.

Water supply in the town

The importance of the Bell Brook and its role in Viroconium are given further emphasis with the eventual construction of the Roman town's water supply. The aqueduct is known to have been a leat sourced by damming this watercourse at a point 1.6 km northeast of the town (White and Barker, 1998: 99). From this point, the conduit appears to have followed the course of the Bell Brook Valley before entering the town close to the East Gate. The date of the aqueduct is unclear. Evidence of the legionary phase at Wroxeter suggests a requirement for water (baths and a possible fountain), meaning it could date from the first century AD. However, what seems sure is that the water supply was at its peak during the second-century building phase of the site. Wacher (1975: 442) asserts that the conduit could have terminated at the site of an assumed Romano-Celtic temple, just north of the baths. He refers to aerial photographs of the site which show a faint outline (possibly a water pipe) entering the area that he tentatively identifies as a sacred enclosure. There have been numerous finds of anatomical *ex-votos* in the form of eyes made of gold, copper alloy and plaster. These were mostly found redeposited in rubble on the baths'

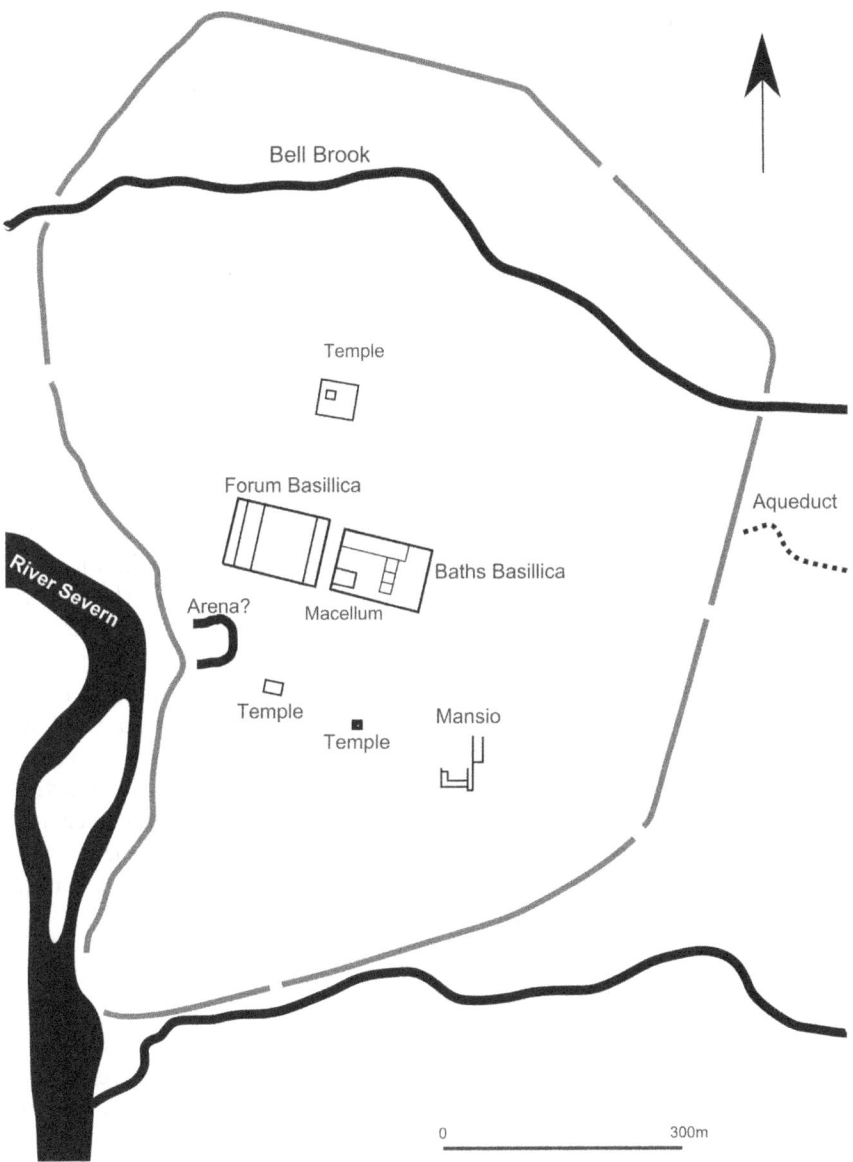

Figure 3.14 Plan of Roman Wroxeter with some of its primary features.
Source: Drawn by author after Rogers (2008).

basilica site, which possibly attests to a temple in the city specialising in the healing of eye complaints (Ferris, 2002: 1). Such activity is often linked to water in various periods, and thus the association with the aqueduct is fitting. Incidentally, in her widespread analysis of Roman wells, Ross (1968) notes three examples at Wroxeter that she deems as being of potential ritual significance. These were found during the Bushe-Fox excavations of 1912–1914, which covered the area just south of the forum. The finds included ox bones, an iron axe, iron knife blades, a pair of bronze tweezers and a number of whole pots (ibid.: 274).

In terms of other internal water supply structures, Bushe-Fox (1916: 13) found a conduit that ran along one of the central streets, with sluice gates providing water for private residences. The other primary beneficiary of the water supply may have been the *macellum* (the issues of interpretation when it comes to British *macella* are noted in the discussion of evidence at St Albans). Discovered in a Hadrianic build-up of material in the area was a fine-grained sandstone sculpture of Venus, or perhaps a water nymph. This figure is regularly mentioned in publications (Frere *et al.*, 1984; Webster, 1988; Henig, 1995; Henig and Webster, 2002), and is believed to be a decorative outlet for a fountain. The *macellum* had a drain and substantial foundations, suggesting that a fountain was placed in the central court (Ellis, 2000: 342). If the water of the aqueduct was already associated with special meaning, its focus within the *macellum* area might have been indicative of further associations of a local nature. There is much evidence to imply increasing ritual associations with food during the Late Iron Age (Bradley, 2000: 152). The idea of wet valley locations being key meeting points in prehistory for connectivity and trade is also noteworthy. The relatively unchanged peripheral rural economy of the Cornovii territory, with its copious examples of pit alignment/dyke waymarkers to watery locations, suggests that such attributes were particularly relevant in the trading centre of the town's *macellum*.

Local activities and the bathhouse

Of course, most of the water supplied to Wroxeter was guided towards the town's bathhouse. Indeed, much of the internal piping that has been found during excavation has related to the bathhouse and its various phases. There was a legionary bathhouse at the site of Viroconium, but this was dismantled during the creation of a forum space and main public bath building in the mid-second century in reign of Hadrian (Webster, 1988: 140; White, 1999: 279). Work commenced circa AD 120 and was completed some 30 years later. It was probably one of the largest baths in the province (Burgers, 2001: 69), and there is a possibility that Hadrian himself had a financial role in the development, due to his known interest in provincial building projects (Barker and Webster, 1990: 2; Duncan-Jones, 1990: 66).

However, it is more likely that the Wroxeter baths were funded by cumulative donations from civic authorities (Millett, 1990: 289). As stated earlier, the line one draws between the central urban authorities and private benefactors is perhaps somewhat blurred. It seems inevitable that prominent individuals from the community would have been linked to the construction. In addition, the extended building time of the project suggests a sense of the baths becoming a communal endeavour rather than a hastily erected monolith for incoming people. That drawn-out process is reminiscent of the creation of some monumental Iron Age features found in this region, mentioned above. It also works with current interpretations that put the focus firmly on a local elite using Viroconium to project their traditional power in a new political climate, rather than an incoming host establishing home comforts.

The temple termination point of the Wroxeter aqueduct hints at a local significance for the water that supplied the baths, and the *ex-votos* suggest a healing function. The further discovery of a *collyrium* stamp in association with the structure increases the likelihood of that interpretation (Jackson, 1999: 110). These small finds are evidence of medicinal ointments being used in the baths, and are often associated with the healing of eye conditions. Traditionally this would be seen as a 'Roman' function for the baths, but Baker (2011) questions whether these artefacts are indicative of incoming Mediterranean medicinal practice. The distribution of *collyrium* stamps is weighted heavily towards temperate European contexts, rather than Italy (ibid.: 159). Moreover, in these provincial settings it is comparatively rare to find them in a Roman military settlement (ibid.: 163). Consequently, while the application of *collyria* was a practice stemming from a Mediterranean context, the proportions witnessed in provincial settings could be symptomatic of a link to a more pervasive local tradition of ocular healing. In line with this, it is noted that while *collyrium* stamps referred to particular types of *collyria*, these have not always been found to match (when both have been recovered). Thus it may be the case that similar local remedies have been matched to the original Latin *collyria* inscription (ibid.: 164). The last aspect to mention regarding the stamps is that they seem to have been deposited in watery contexts. In Gaul a significant number of these items are found in rivers, wells, pits and baths; these could be underrepresented, since the artefacts are small and easily missed in river dredging (ibid.: 171). This appears to link the stamps to the well-observed water traditions of the northern European Iron Age, and therefore casts a different light on the example found at Wroxeter.

Further diversification of activity is found in the Late Antique phases of the structure. In this period, metalworking and burials occur in the baths and associated basilica (Wright, 1872: 68; Barker and Armour-Chelu, 1997: 72–79). The subsequent interpretation of an early church being built in the *frigidarium* area only serves to further the ritualistic link (Rogers,

2008: 142). Regardless of its specific use, the evidence of timber structures surrounding this central building in the fifth century implies its continued prominence (Barker and Armour-Chelu, 1997: 138–168). This enduring quality, together with the evidence of diverse activities, suggests that the Wroxeter bathhouse represents far more than the simple adoption of Roman-style bathing. It became the heart of the community when other 'Roman' features like the forum were falling out of use, so it is plausible to suggest that local adaption and interpretation were at the heart of its continued prominence.

Leicester (Ratae Corieltauvorum)

The waterscape context of Leicester

The Roman town of Ratae Corieltauvorum was constructed at a dynamic point on the River Soar where it braids to form an island. There is a suggestion that on this small strip of land, between the two arms of the river, an early Roman fortlet was established (Clay and Pollard, 1994). Evidence of this is limited to a length of military-style ditch, and is not necessarily conclusive proof of sustained occupation of this area by the Roman army; instead, it may be defining a significant enclosure of some sort (Cooper and Buckley, 2003: 33). Regardless, the bridging of the river at this point in the Roman period formed the western entrance to the settlement, and much of the monumental infrastructure of the town was established nearby. However, underlying these Roman developments is considerable evidence of Iron Age occupation. Early excavations at the Jewry Wall site uncovered significant pre-Roman pottery (Kenyon, 1948: 24), and at Bath Lane, St Nicholas and in the West Bridge area of town other elements of settlement were found, such as fragments of flan trays used in preparation of Iron Age coin blanks (Clay and Pollard, 1994: 72; Cooper and Buckley, 2003: 31). Evidence of Iron Age pits and gullies in the St Nicholas Circle area and at 51 Thornton Lane suggests the presence of roundhouses (Rogers, 2008: Diagram 32). Some finds (especially the imported pottery) demonstrate that the settlement already had exceptional importance in the Late Iron Age, and so its progression into a Roman *civitas* capital was a natural evolution. The central St Nicholas Circle area eventually became the monumental heart of the Roman town; the forum and the Jewry Wall baths are both notable features (Figure 3.15).

In terms of potential religious associations with this space close to the river, one confirmed temple has been discovered (Wacher, 1995: 359). As with a number of ambiguous structures mentioned in this chapter, it has been loosely interpreted as a *mithraeum*, but many have been unconvinced by this interpretation and feel it could equally be a shrine to another deity (Henig, 1984: 95). The nearby recovery of a figure that appears to be a

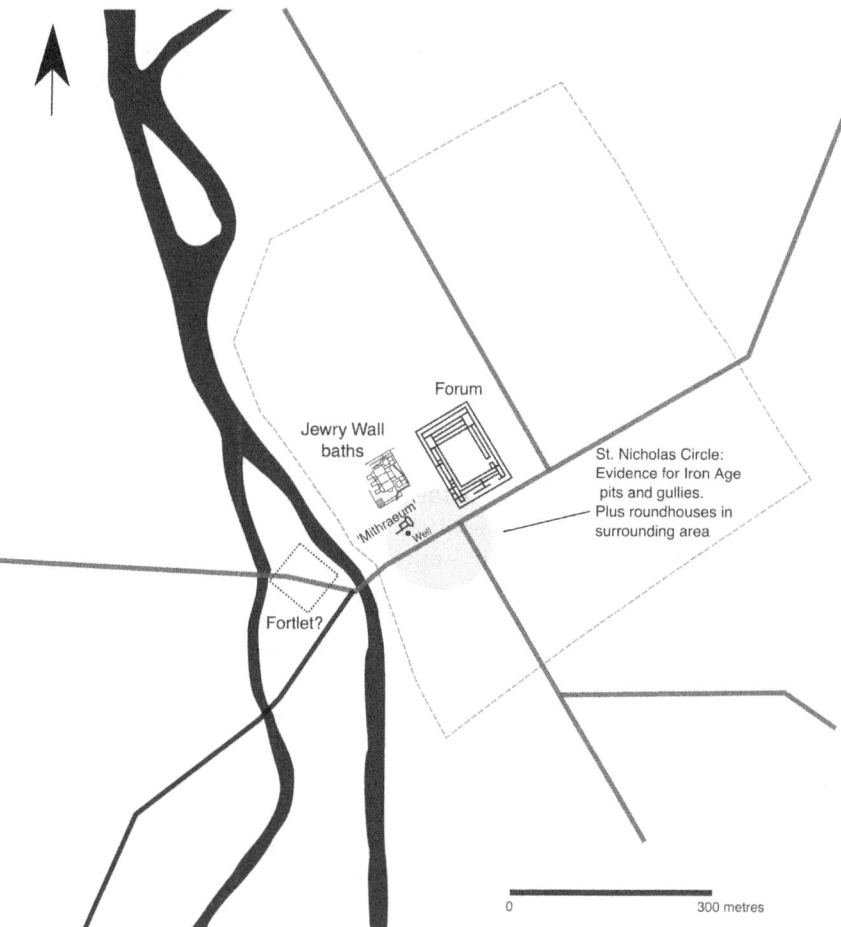

Figure 3.15 Plan of Roman Leicester with the notable features of St Nicholas
Circle.

Source: Drawn by author after Cooper and Buckley (2003).

water deity, associated with a frieze or pediment ornamented with sea ser-
pents (Wacher, 1995: 359), plus an unusual cult figurine of Mercury combined with a local god could indicate equally viable options (Pollard,
1998: 355).

Immediately outside and to the south of this building a well was found,
but it has been largely ignored in the analysis of the immediate area
(ibid.: 353). Wacher (1995: 359) makes a note of the feature but does not
elaborate any further. This is remarkable, as it produced an exceptional
collection of pottery with a high proportion of beakers, many of which

were virtually complete (Pollard, 1998: 353). The assemblage was mostly of a third-century date, but that does not preclude earlier significance for this feature or the space where it was constructed. The proximity to the temple means it is entirely reasonable to envisage the two being linked. The large sample of beakers may suggest that the act of pouring water, perhaps into the well, or transferring it to the temple could have been a particular focus. While it is difficult to make connections between Iron Age uses of space and these third-century developments, it may suggest some sort of continuity of importance for the area connected to water. Regardless, it is certainly another example of a well located in a prominent place in a town, with more than just practical associations.

The Raw Dykes as a possible aqueduct

For many years archaeologists have debated whether the Raw Dykes monument (to the south of Leicester) indicates the remains of an aqueduct. Kenyon (1948: 41) put forward the idea after finding a few sherds of Roman pottery in association with the feature. This was combined with apparent confirmation from eighteenth-century sources suggesting the Raw Dykes had once stretched towards the city. If it were an aqueduct, the only source nearby would have been Knighton Brook, although this is a rather small watercourse in the modern day (ibid.) (Figure 3.16). However, Kenyon was writing in the context of trying to find an ample water supply for the impressive Jewry Wall bathhouse. As such, this interpretation of the Raw Dykes was still problematic. She noted how the baths were actually at a higher level to the supposed aqueduct, and thus would have been impossible to supply without a lifting mechanism. As a result, Kenyon (ibid.) envisioned the Raw Dykes as representative of work by an 'incompetent provincial engineer' which never succeeded in bringing water to the town.

The case of Leicester is so puzzling because of the extensive drainage features that have been associated with the town bathhouse. Wacher (1995: 349) notes that major drains ran around at least three sides of the structure, implying plenty of water in the vicinity. Burgers (2001: 42) put forward the idea that water could have been supplied to a central area of the town before being lifted by a tower structure; this has precedent in the Mediterranean. Nevertheless, there is sparse evidence to corroborate the existence of such a structure in Leicester. Wacher (1995: 350) mentions the discovery of a small ditch (1.2 metres wide and 1.2 metres deep) running north away from the approximate point where the supposed aqueduct would have entered the town. This ditch could be the traces of distribution pipes from a terminal point close the South Gate. He also mentions a stone-basin drinking fountain found in 1862 on the site of 52 High Cross Street (ibid.); this would have been on the principal

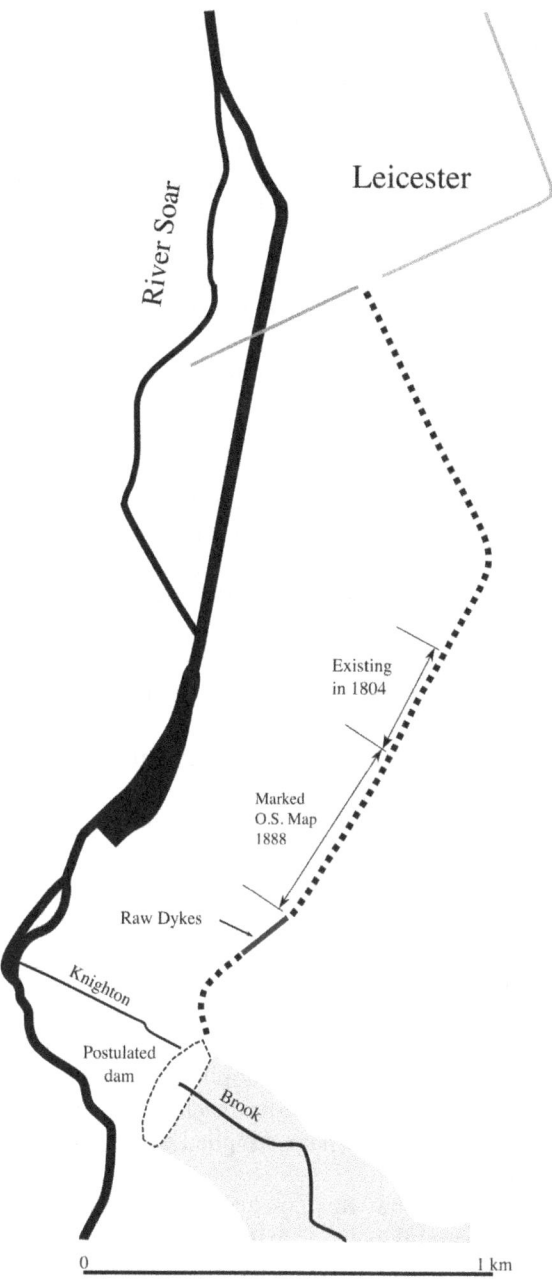

Figure 3.16 A suggested course for the possible Raw Dykes aqueduct.
Source: Drawn by author after Wacher (1995).

north–south street of the town. This implies the existence of a relatively advanced running water network, but does not aid the interpretation of the Raw Dykes. Indeed, Burgers (2001: 42) stresses how more recent excavations between the town and the monument have not found any trace of its continuation towards the Roman settlement.

The fact remains that if the Raw Dykes was an aqueduct, it was not necessarily a requirement for it to supply the bathhouse. If a conduit reached the gate it could still have a profound symbolic impact on the entrance to the settlement, and may have served the immediate buildings in this area. In line with this, Wacher's (1995) presentation of the assumed path of the conduit, from the Knighton Brook to the town gate, raises some interesting questions about its purpose as a landscape feature. As with the example at Dorchester, the conduit mirrors the nearby river, which in this area had a tendency to braid. Leicester's Late Iron Age activity centres on the Soar itself, moving away from previous settlements in higher areas (Charles *et al.*, 2000; Cooper and Buckley, 2003). Some of the first 'Roman'-era activity on the site seemed to mark out a gravel area underneath the later forum, which could have been the crystallisation of meeting points close to the river during prehistory (Wacher, 1995; Cooper and Buckley, 2003). Consequently, it makes sense that the river may have represented a key point of movement through the valley from the surrounding territories. Navigating the marginal areas around the Soar could subsequently have constituted a common prelude to meeting in the central areas that later came to form the Roman town.

The importance of marking such areas may have been part of the reason for monumental earthworks in prehistory; the derivation of the name 'Ratae' has been linked to the construction of earthworks in the area (Rivet and Smith, 1979: 443). Although it is now seen as a failed water supply structure, the Raw Dykes was formerly seen as a prehistoric construction, serving a similar purpose to dykes outside settlements such as Colchester. However, a Roman-period aqueduct does not necessarily have to be divorced from these associations: it may have served to highlight the central importance of Leicester in the landscape, and also to delineate the zone of meaning close to the river. Importantly, the Raw Dykes does not necessarily have to have been an aqueduct that supplied a whole town in a practical, empirical sense. Other sources of water (such as the Soar itself), or an as-yet-undiscovered conduit, could have supplied the bathhouse and the higher levels of the town.

Likewise, it is entirely possible that this construction did not have the practical supply of the *civitas* capital as its primary aim. It could have been more concerned with defining a sense of place that was forming in the Late Iron Age and early Roman period. The entranceways in the Iron Age settlements of this region are known to have been unique places, worthy of respect, with evidence apparent even in the immediate surroundings of

Ratae (Cooper *et al.*, 2006). Thus the association of a dyke with an entrance is not something totally out of character in the prehistoric landscape. There is also a possibility that the structure could have taken water away from the town, as is suggested of the Lincoln aqueduct. If future survey and archaeological work cannot prove more decisively the purpose of the Raw Dykes, perhaps these phenomenological lines of enquiry represent the best chance of understanding the role of this ambiguous feature in the Roman period.

The Jewry Wall baths

Regardless of whether the Raw Dykes was the source of water for central areas of the town, the 'Jewry Wall' in modern Leicester is a testament to the impressive size of the bathhouse that once stood in the centre of the town. The creation of these baths was part of second-century building projects that included the nearby forum (see Kenyon, 1948). The layout of the building was quite unusual for a bathhouse within the province, in that it did not have a sequential series of rooms providing increasing heat (Wacher, 1995: 349). Instead, the *caldaria, tepidaria* (three of each) and *frigidarium* (single) were created in a symmetrical lateral spread of three *caldaria* and *tepidaria* next to each other (Burgers, 2001: 75). It seems that the main concourse of the building served as the *frigidarium* and thus probably had basins of cold water throughout (Wacher, 1995: 349). This layout makes it hard to describe convincingly the way one experienced the building and the sort of activities that were taking place. The symmetrical layout could have allowed room for separation of social classes, or indeed particular social and business activities. As a result, one has to consider the ideas expressed above: that local custom could have had a greater than anticipated impact on the conception of the building.

We must thus return to the other activity in the insula referring back to the temple and well (Pollard, 1998), and judge the Jewry Wall baths of Leicester in this immediate context. If the local concerns of the bathhouse are considered, the potentially unique nature of the water being utilised could have played a prominent role in how people conceived the structure. In this regard, the drains of the baths could be particularly relevant. Wacher (1995: 349) notes that the examples found in association with the Jewry Wall baths were incredibly generous when traced in the south and southwest directions (two channels). In fact, they were so massive in size that there was sufficient headroom for a man to walk down them (ibid.). These drains are often used to support the assumption that the town had a significant water supply. However, the direction of the drains to the southwest and south would tie this bathhouse into the area of the so-called *mithraeum* and possibly the crossing point of the Soar. The grand nature of the structures infers a large amount of water flowing into (or out of?)

these areas. The potential for heightened significance for these baths is supported somewhat by their role in Late Antiquity (and beyond). Even today, a church (St Nicholas) stands next to the Jewry Wall remains, and it appears that the bathhouse was incorporated into a Christian area'of worship before this Anglo-Saxon building was constructed (Kenyon, 1948: 34).

Specifics of the baths aside, what seems clear is that the development of the central monumental area of Roman Leicester was fundamentally linked to the movement of water. Further archaeological work may help us understand more about this relationship, but it should not be surprising considering this was the *civitas* capital of a region that encompassed the water-centric settlement at Lincoln, and was inhabited by an Iron Age tribe known for their links to rivers (see Breeze, 2002).

Colchester (Camulodunum/Colonia Claudia Victricensis)

The Iron Age waterscape of Colchester

As the first capital of Roman Britain and centre of the imperial cult, Colchester was a settlement of primary ideological focus (Wacher, 1995: 118). This significance did not suddenly appear in the Roman period: principally from the Late Iron Age onwards this was an important site, possibly reaching its peak of influence during the reign of the Catevellauni leader Cunobelin. Increasingly, scholars have looked at this Late Iron Age ruler as an equivalent to the client kings elsewhere in the Empire (notably Asia Minor), whereby he was deeply familiar with Roman practice and symbolism because of direct experience in Rome as a child or adolescent. This is seen in both his coinage (Creighton, 2000) and the layout of nearby Gosbecks, with Cunobelin mimicking the Roman fort style (Creighton, 2001: 8). Of course, Roman-period buildings such as the Temple of Claudius appear to be features planned by the central imperial authority, but the familiarity the rulers of this region had with Rome does place urban development within a dynamic context. These factors are inconsistent with seeing Roman Colchester in black-and-white terms of 'Roman' and 'native'.

The immediate topography of Colchester is another example of a settlement woven into a wetland context (Figure 3.17). The town is located on a peninsula formed between the confluence of the Colne and Roman Rivers (Hawkes and Hull, 1947: 3). Downriver from Colchester the Colne empties into the sea, and in the estuary area Mersea Island was a focal point for prehistoric burial.[9] The Iron Age *oppidum* at the Colchester site stretched over this landscape, with what archaeologists have deemed an important industrial site at Sheepen (Hawkes and Hull, 1947; Niblett,

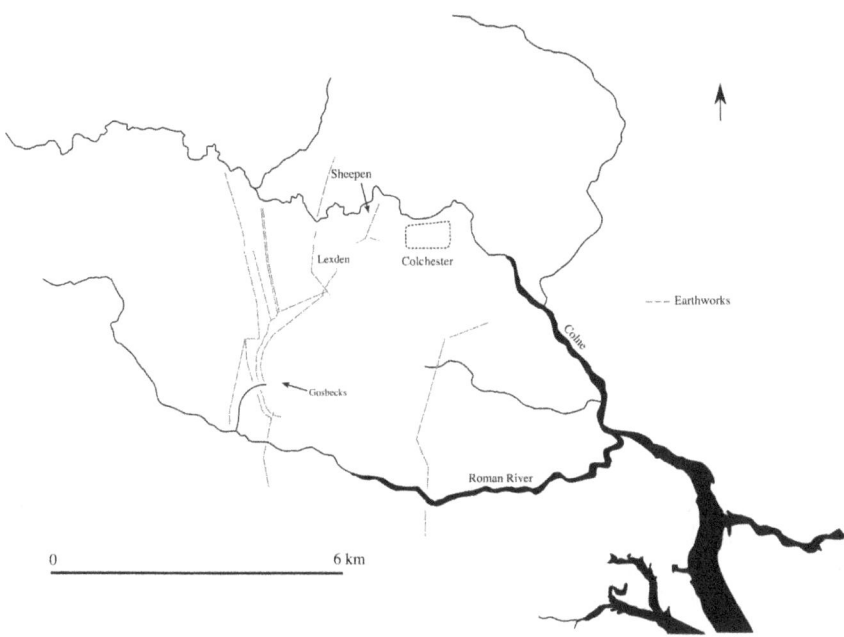

Figure 3.17 The wider setting of Roman Colchester, with its prominent earthwork features.

Source: Drawn by author after Hawkes and Crummy (1995).

1985; Crummy, 1997), some 0.75 km northwest of the Roman town
centre. Willis (2007b: 121) highlights how the Colne was particularly
dynamic at this point, where several streams drain into the main river. This
means the floodplain was wide, with potential for areas of standing water
(Rogers, 2008: 78). The Sheepen site aligns with the lowest non-tidal part
of the river, indicating a transition in the waterscape. Indeed, Willis
(2007b: 121) mentions that all these elements would create a landscape
rich with religious and symbolic meaning; this is something reinforced by
the proliferation of temples constructed here after the Boudican revolt
(Hull, 1958; Crummy, 1997). Sheepen also served as a fording point of the
river (Hawkes and Hull, 1947: 4), which in itself perhaps adds to the
symbolism of the area.

At Colchester we can see evidence of another large-scale system of Iron
Age earthworks. Again, there can be tentative suggestions about a relationship with water – many of them, for instance, start or finish at either the
Colne or the Roman River. Garland (2017: 138) argues that these earthworks were communal aspects of the Late Iron Age that clarified the
meaning of the watery landscapes which surrounded Colchester. Part of

the system established in the Iron Age also defines the Sheepen site. One pronounced pattern worth mentioning in this regard is found in the dykes around the Iron Age and Roman ceremonial site at Gosbecks – a place that has clear religious value in the Roman period, with a temple complex, theatre and a notable source of spring water (Rogers, 2008). All the surrounding earthworks of this site leave a clear gap for the stream that runs from the heart of the settlement. This seems to monumentalise the flow of water within the immediate landscape. The effect of the dykes could also have been to direct more surface water down the valley into the Roman River.

Roman water supply at Colchester

Colchester's human engagement with water is clearly evidenced in the Iron Age, and appears to have intensified in the Roman period. The construction of Roman bridges at specific points may have been an early and symbolically relevant undertaking. Hawkes and Hull (1947: 4) note that the high-tide mark of the Colne was located opposite the Roman East Gate, where it is presumed the eastern Roman road crossed the river. The northern entrance to the town is preceded by another crossing, which could also reference a significant point in the river. After this the Colne curves northwards and meanders into the Sheepen area; the Roman crossing possibly acknowledges this change in the river's nature.

For water supply, the town had at least five springs within its walls (Crummy, 1984: 27; Burgers, 2001: 98). However, unlike in other towns outlined earlier, no definite examples of Roman wells have been found within the central area of settlement (Crummy, 1984: 26) (see Figure 3.18). In part this is because the town lies on a 50-metre contour, which means obtaining groundwater in the central areas could have been a major undertaking. This obstacle does not necessarily preclude Roman well digging: the evidence from the medieval period shows it was certainly possible (ibid.), and Burgers (2001: 58) rightly notes that large areas of the modern town could still conceal wells that would enhance the overall picture of water supply. But the recent change in archaeological focus in Colchester, to the periphery of the Roman urban area (see Crummy, 2005), may result in this picture remaining unclear for the foreseeable future.

There are some instances of wells outside the central Roman town. Examples are noted at Middleborough, between the North Gate and the Colne; Sheepen, a site already mentioned as having ritual significance in the Iron Age and Roman periods (Hawkes and Hull, 1947: 53); and Chiswell, close to the road leading to the Balkerne Gate (Crummy, 1984: 26). The Sheepen wells are interesting because of the ritual interpretations of the site (Willis, 2007b; Garland, 2017) and the presence of at least four temples built in the Roman period (Hull, 1958). Traditionally

Figure 3.18 The known urban water features of Roman Colchester.

Source: Drawn by author after Burgers (2001).

the wells have been linked to proposed industrial activities happening at the site, but it is possible that our developing knowledge of the ritual life of the area may alter this interpretation.

The evidence at Middleborough provides us with an intermediary between this area of prehistoric significance and the established Roman town, in a location that seems to have been consumed by the floodplain of the Colne with areas marked by standing water (Willis, 2007b: 121). The three wells discovered in Middleborough were located close to the road which left the North Gate en route to Sheepen itself. Despite not containing any remarkable finds (Brooks and Crummy, 1984: 182), it is possible they had a degree of spatial importance, potentially being involved in processional activities that took place between the main town and the places of prehistoric importance in the immediate landscape. The (largely accepted) proposal of a processional link between the Iron Age site of Gosbecks and the Roman-period town (Esmonde Cleary, 2005) is an example of this aspect of movement. With our changing understanding of Sheepen, the route from the North Gate may have been less about commercial transport and more about individuals coming towards the site with ritual goals in mind.

Moving away from wells, there is slightly more substantial evidence for aqueduct water being harnessed from the surrounding landscape. While nothing has been found on the scale of Lincoln, a smaller conduit has been discovered. To the west a 'Claudian Leat' was found running from Sheepen Spring towards Colchester. This appears to follow the natural contours of the land, and may be linked to the significant number of Roman water-mains that have been found close to the Balkerne Gate (Crummy, 1984: 115–117). Any supply reaching this area from Sheepen may have been supplemented by water from nearby Chiswell Meadow, a place heavily exploited for this purpose in later periods (ibid.: 27).

In the early Roman period the Balkerne Gate was the western entrance to the *colonia*. It was undoubtedly an impressive spectacle to those travelling on the main road from London (Isserlin, 1998: 126).[10] In addition to its architectural scale, it was located on high ground with views down into the valley of the Colne (Crummy, 1984: 123). This was an area of particular importance, evidenced by the presence of multiple temples, including two directly outside the gate. A bronze figurine of Mercury was recovered nearby, and provides a tangible example of worship at the site (Crummy, 2006). As already noted, Mercury is often connected to places of liminality, and the combination of a watery setting and a gateway would make an ideal location. Of course, the movement of water from Sheepen to Colchester may itself have borne potent local ritual meaning. The leat actually passes through Sheepen Dyke, and was thus actively engaging with the prehistoric monumentalised landscape of the wider Colchester area.

The other notable water conduit in the town is located opposite the Claudian temple. A pipe led water from the Insula XV '*mithraeum*' structure, towards the northeast gate. While this building is no longer seen as a temple, it is nonetheless constructed on one of the many springs in the town. The idea of channelling spring water out of the town obviously seems counterintuitive. Some have said that the abundance of water from this spring source led to flooding of the immediate area, so channelling the water out of the town has been explained as making practical sense (Hanson, 1971). This interpretation highlights the numerous other water outlets that joined to this pipeline, effectively making this a primary drain. However, it seems strange that a surplus of spring water would not first be channelled to other areas of the town. The fact that the '*mithraeum*' lies opposite the Temple of Claudius adds conviction to the idea that the conduit had a more significant purpose. It is possible that this channelling of water from the symbolic heart of the *colonia* is an acknowledgement of the importance of the Colne. It may also be creating a similar connection to water at the northeastern gate as to that we see at the Balkerne Gate, which could be connected to the particular properties of the Colne at this point.

Locating and interpreting bathing structures

The evidence of bathhouses at Colchester is fragmentary at best. Initial discoveries during a small excavation at 61–62 High Street uncovered what appeared to be a cavity formed by a massive hypocaust (Crummy,

1988: 37). In some ways a public bathhouse in this location (the northeast of Insula XXX) would make sense, as it was a central area directly opposite the Temple of Claudius, but not enough evidence was recovered to enable concrete interpretations. A slightly better-evidenced bath-related structure has been found in the northwestern corner of the town (Insula I): the remains of a room with a tessellated floor, benches and *in situ* wooden piping were a surprising discovery beneath the modern sixth-form college (Brooks *et al.*, 2009). One of the most intriguing elements of this structure is the possibility of a pronounced religious element to its function. The wall decoration of the room during its first phase was white with a floral motif, before changing to a dominantly red interior. The original colours and general positioning of the room (on a slope, close to water features) hint at similarities to *nymphaeum* structures in Italy (ibid.: 35). The change of decor could signify the incorporation of this religious structure into a grander superstructure of baths.

However, the most convincing evidence has been discovered in the east of Insula XX during excavations at East Stockwell Street, immediately east of the centre of the modern town and town hall and a central location in the Roman city. The excavations confirmed the presence of a very large structure constructed in the second century (Benfield and Garrod, 1992: 28). The foundations were around 1.5 metres wide and nearly 4 metres deep below the Roman floor level, meaning it was unlikely to have been a residential building (Crummy, 1991: 9). Ten rooms were uncovered, and many were surprisingly small (considering the foundations); one was in notable contrast, measuring approximately 10.5 metres wide by 27.5 metres long (ibid.). This general area of the Roman city had seen little opportunity for excavation in the past, and hence had been thought likely to contain some of the key civic structures that had to date proved elusive in Colchester: the forum, basilica and main baths. The unusual fragmentary plan revealed at East Stockwell Street eliminates many public building types from consideration: apart from the bathhouse, it could only plausibly have been a basilica, market or some form of palace (ibid.: 8). This interpretation was strengthened by the discovery of a substantial drain in the large room (Benfield and Garrod, 1992: 30).

Part of the problem of deciphering the purpose of the building is that it does not have a traditional plan, even for a bathhouse. This could merely be a product of the partial excavations within the present urban context, and there is certainly precedent for a central drainage feature in the *frigidarium* of British bathhouses (Zienkiewicz and Allen, 1986: 60–65). And the distinctiveness of this plan could be just as influenced by some form of local variability and purpose for the bathhouse itself. It is possible that the basilica-type form of the *frigidarium* is informing us of an enhanced social aspect involved in the act of bathing within Colchester; this could feasibly be linked to attitudes towards the medium of water. However, Colchester's modern circumstances limit the extent to which we can understand connections between the various water features of the central town, and this limits the extent to which this prominent site can help us understand Romano-British urban water engagement. Nonetheless, there is enough evidence to state that in the landscape of Colchester water was clearly a marked concern in both the Iron Age and the Roman period.

Chichester (Noviomagus)

Water, the Chichester Entrenchments and the Downs

As is the case at St Albans and Silchester, the most noticeable remaining vestiges of the prehistoric landscape of Chichester are a series of earthworks, known as the Chichester Entrenchments (Figure 3.19). These features have been traditionally conceived as structures relating to the defence of either a local *oppidum* of the Late Iron Age or possibly the important 'palace' at Fishbourne (Cunliffe, 1971: 15; Bedwin and Orton, 1984). In part this is due to their rectangular arrangement, seemingly protecting the area of Chichester. This is slightly unusual, because such features do not

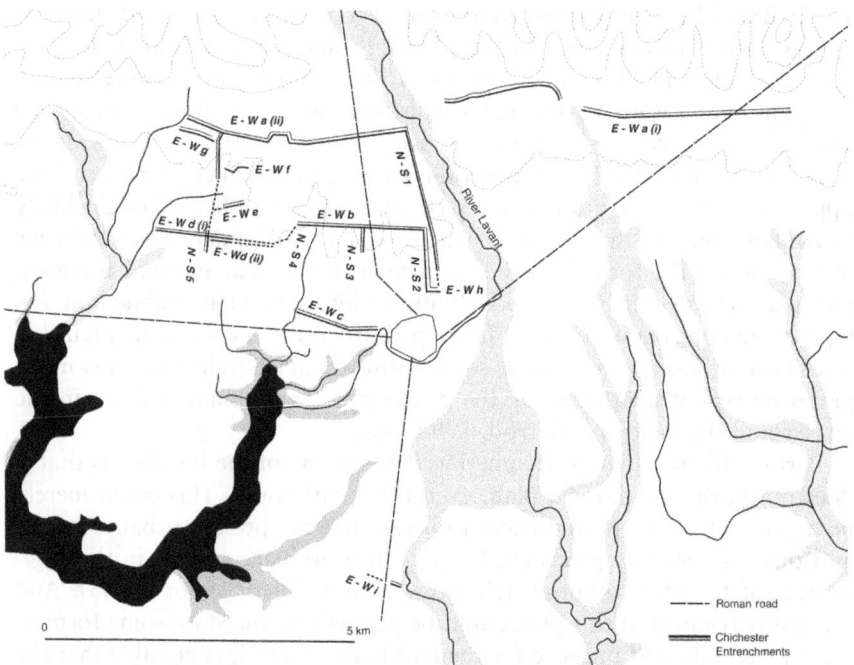

Figure 3.19 The Chichester Entrenchments and the surrounding waterscape.
Source: Drawn by author after Bradley (1971).

often present as a comfortable defensive enclosure area (Rogers, 2008: 70). Bradley (1971) confirms that it is somewhat unrealistic to view the earth-works as a unified whole; rather, there were a number of stages that built up to this convenient arrangement. Despite this, he conforms to the general ideas of Frere (1967: 46) that the purpose of the Chichester Entrenchments was likely to have been defensive and also to delineate large tracts of settled land (Bradley, 1971: 34). This idea of dyke systems creating a sense of ownership or dictating power relationships through movement was explored by Witcher (1998) from a phenomenological perspective. Yet one of the most interesting elements of Bradley's survey is that it highlighted a deep relationship between the Entrenchments and the surrounding water-scape. In fact, each of the outlined phases of development is closely related to the pattern of rivers and the surface geology (Bradley, 1971: 30).

The earthwork labelled E-Wa (ii) in Figure 3.19 is the clearest example of such a relationship, running between the River Lavant and Bosham watercourse. Likewise, E-Wc ran between the Lavant and a watercourse that discharged into Fishbourne Harbour (Bradley, 1971: 24); E-Wa (i) seems to respect the line of a major watercourse while also running through a wet floodplain area (ibid.: 20); E-Wd runs east from a stream that discharges at Bosham (ibid.: 23); N-S1 starts close to the Lavant before heading south towards what would become the Roman town of Chichester; and there is even the possibility of E-WI representing a small dyke running from a watercourse south of Chichester (Curwen, 1954). In addition, the dyke systems highlight at least two springs in the surround-ings of Chichester. The course of E-Wa (ii) leaves its alignment at one point, forming a rough trapezium shape; there is evidence to suggest that a spring was located within this 'salient' (Bradley, 1971: 21). Furthermore, N-S5 could have linked the two north–south dykes at what was described in antiquarian reports as 'The Watery Line', where a spring was located (ibid.: 24). Both examples show significant points of deviation for the Entrenchments, and thus should perhaps be seen as of equal importance to the termination points at rivers. There is also a possibility that the area had many similar points, considering the number of smaller streams surround-ing Chichester.

The link between water and these features has only really been explained as beneficial ways to limit movement or mark out space in the landscape. However, as with the examples at St Albans, there is a possib-ility that we should analyse these dykes as a conscious effort to manipul-ate or highlight the local waterscape. Periodic standing water (from rainfall) is possible. The proximity of springs creates the possibility of consistent direct water drainage into the Entrenchments. The two spring sites highlighted above may well have also had 'entrance' points to the earth-works (ibid.: 33). In this regard, the north–south Entrenchments are of particular interest because they run from the high to the low ground, from the direction of the Downs to Chichester. It is debatable whether this could have created any consistent flow of water, but the dykes do follow a similar path to that of the Lavant. The relationship between the lowland coastal location of Chichester/Fishbourne and the South Downs to the north could have been particularly important. The River Lavant, for instance, is a product of the chalk composition of the higher ground. While it is known as a winterbourne river today, this may be a result of high extraction rates by the local water authority (Rogers, 2013: 95), and it could have run far more consistently and powerfully in the Roman period. The characteristics of such chalk rivers may also lend themselves to further meaning. They are an almost uniquely British phenomenon, with 80 per cent of the world's chalk rivers occurring in England (WWF, 2014). Filtration and storage in the chalk mean these watercourses retain a relatively high temperature and a clarity not always apparent in other rivers. These make them ideal for drinking water, but they also serve as pivotal sources of biodiversity. In addition, they can produce unusual visual displays in the landscape. Due to the relatively high water temperature, on cold winter days these streams can appear to have steam rising from them. These attributes could all combine to make the Lavant a particularly symbolically rich watercourse which may have forged a deep connection between its source in the South Downs and Chichester.

That notwithstanding, the Downs are also well known for their dry valleys or 'coombes'. These dramatic landscape features, such as the Devil's Dyke north of Brighton (where a hillfort overlooks and is half enclosed by this landform), may be dry now but their explanation always comes back to the presence of water. It is probable that they were conceived as remnants of powerful rivers that once flowed through the landscape (Tilley, 2010: 71). The question of where this water went would surely have been a part of local legend, and might have involved an acknowledgement of its presence buried in the land of the Downs. Additionally, these dry valleys are like colossal versions of the Entrenchment system, hence the human-made dykes around Chichester could be partly referencing the water associations of the higher ground. The fact that the E-Wa (i) entrenchment closely follows the edge of the downland chalk (Bradley, 1971: 21) could be further evidence to illustrate this link.

Water and Noviomagus

Noviomagus provides us with one of the best epigraphic sources displaying the veneration of water deities in Roman Britain. The dedication of a temple to Neptune (*RIB*: 91) seems to communicate an accepted heightened value of water. Unfortunately, the evidence for an external water supply is not particularly forthcoming. Wacher (1995: 264) notes how the antiquarian sources speak of a conduit entering at the North Gate. This is interesting, because these reports could well link to the earthworks: the north–south Entrenchments were in the area of the North Gate of Chichester. It is possible that the spring sources outlined previously were used in the Roman period. This suggests a degree of negotiation with these prehistoric monuments, perhaps utilising them for the deliberate supply of water.

A more reliable source of water may have been the Lavant itself. The aforementioned origin of the watercourse in the uplands chalk would make it a particularly good source of water for consumption, and also any type of display. Bradley (1971: 30) notes how the course of the Lavant departs from its original floodplain when it meets Roman Stane Street. It then follows a course around the town before discharging into another stream close to Chichester Harbour and the present city wall. This conspicuous alignment to Roman-era structures is the basis for attributing it to this period. However, there is very little evidence to date this activity firmly, and there have been many conflicting proposals (Down, 1988; Magilton, 1996; Rogers, 2013). In the context of the Chichester Entrenchments outlined above, this development would undeniably strengthen the connections between the settlement and its waterscape. Not only is another river-to-river connection created by this diversion, but the water-based link to the Downs is completed.

Regardless of the alteration of the Lavant, the town had additional water provided by an unusual stone cistern that penetrated deep into the water table. It was massively constructed of large squared greensand blocks, and the bottom was around 3 metres below the contemporary ground level (Down and Bayley, 1978: 149). The water was pumped up into the tank manually before presumably being channelled through pipes to the nearby baths (Frere and Tomlin, 1991: 290; Wacher, 1995: 264). In many ways this represents an internal aqueduct; the use of pressure would appear to differentiate it from many well features we find. Clearly, without a sense of wider distribution in the town it is difficult to say how significant this feature was in the settlement. It certainly seems to have provided a supply for the public bathhouse, which was located in the northwest quadrant of the assumed Roman town plan (Figure 3.20). This whole structure covered approximately 5,500 square metres. Construction appears to have commenced in the first century, with a possible Flavian date for an initial working structure (Down, 1978: 145), so there is a high probability that the native ruler Cogidubnus played a significant part in the building of the baths. This is interesting, because the Temple of Neptune and Minerva, with its dedication to a water deity stone-cut in Purbeck marble (*RIB*: 91; Bogaers, 1979), was established by the authority of Cogidubnus and is another feature of the northwest quadrant. Too often this has been interpreted as a native ruler striving to produce a more Roman urban form, but it could be a more nuanced expression of local significance attributed to water.

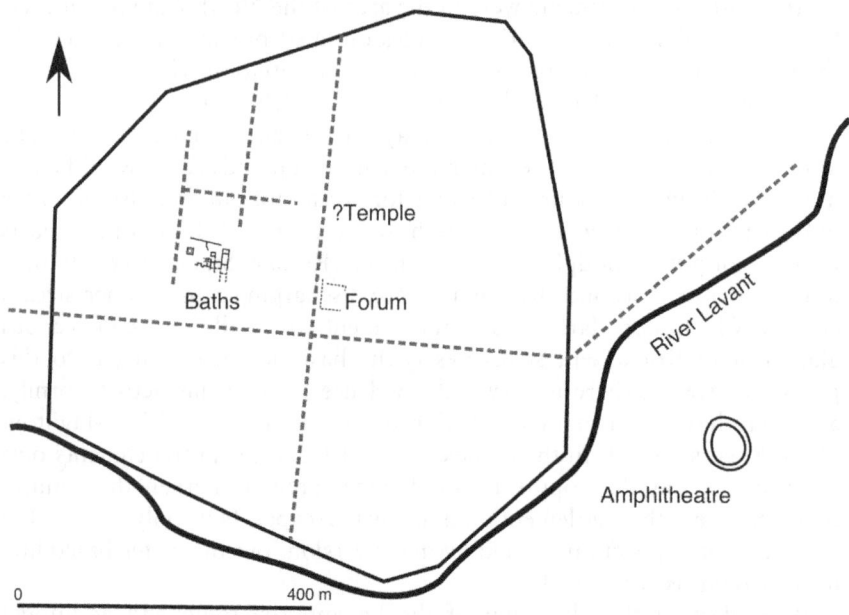

Figure 3.20 Plan of Roman Chichester with the possible course of the River Lavant.

Source: Drawn by author after Down (1978).

Supporting this, Garland (2017: 200) suggests that processions could have started in this area of the town and been directed westwards to important watery focal points in the surrounding landscape. For instance, using the distant tower of the Hayling Island Iron Age ritual complex as a marker, individuals could move westward towards the Ratham Mill temple site. The structures at Ratham Mill were located next to the Bosham stream, and there may have been a central well at the heart of the small temple (King and Soffe, 1983). This is a compelling interpretation because, as mentioned above, parts of the Chichester Entrenchments earthwork system appear to highlight this westward movement through the land-scape, and the Bosham stream is a termination point for more than one of the earthworks. The combination of these factors goes some way to high-lighting the prominent communal role water seems to have played in this settlement; this may also be reflected in the fact that the public baths were one of the longest-serving buildings of the town (Down, 1988: 152).

Winchester (Venta Belgarum)

The Itchen, Oram's Arbour and the Roman aqueduct

Venta Belgarum is worth covering in this chapter, because while it lacks some of the definitive evidence of central monumental buildings, it was deliberately located in a watery setting, being built in the extensive flood-plain of the Itchen over a tufa island. With the town's location close to the Downs, much of what is said in the previous section on Chichester could be repeated here. This includes the fact that the Itchen is another notable example of a chalk river. Unlike at Chichester, there is far more definitive evidence of an aqueduct at Winchester. It is thought to have started at Itchen Stoke Springs before following the course of the nearby river towards the town (see Figure 3.21). This path appears relatively unremark-able, but is actually associated with numerous places of prehistoric signifi-cance. Along the course, for example, Itchen Stoke, Itchen Abbas and King's Worthy all have some evidence of settlement before the Roman period. In addition, the apparent departure from the river course around Woodham's Farm and King's Worthy is notable for its banjo enclosure (Perry, 1966), and the area near Headbourne Worthy (where the aqueduct veered towards its westerly extreme) is the site of Flowerdown Barrow cemetery. The conduit seemingly entered the town close to the North Gate,

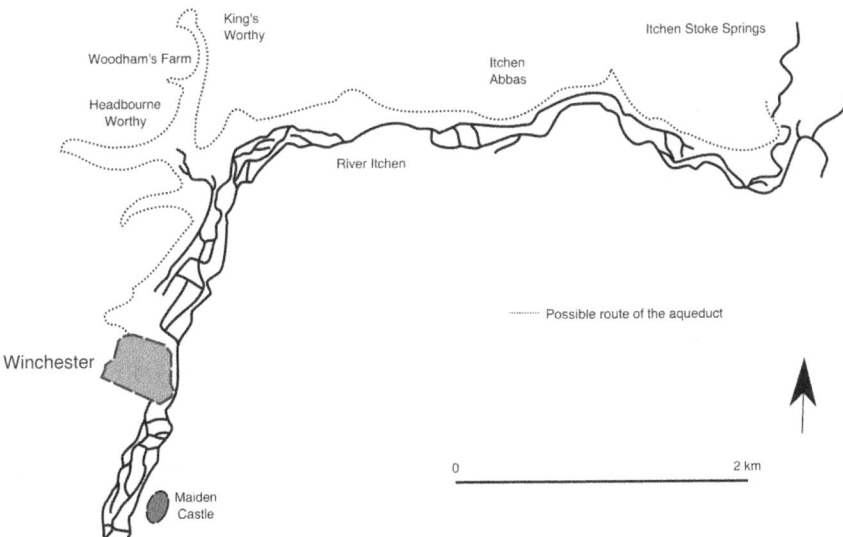

Figure 3.21 The course of Winchester's aqueduct through the surrounding landscape.

Source: Drawn by author after Burgers (2001).

and thus would have had the requisite height to distribute waters to all parts of the town (Burgers, 2001: 43).

Recent work just south of this presumed gateway seems to confirm the potential of this interpretation. Excavations at the Discovery Centre site unearthed a channel 35 metres long associated with a major Roman street. This consisted of a steep cut of 2.5 metres in width, the sides of which were faced with three courses of roughly worked flint nodules bonded by a hard buff lime mortar (Ford and Teague, 2011: 54). Measurements along the base of the channel suggest a gentle fall in height coming from the north. The general direction of this feature is from the northeast to the southwest, which it follows for around 10 metres before abruptly crossing the street mentioned above (see Figure 3.22). This change in direction casts some doubt on the interpretation of the feature as being an aqueduct. Part of the problem is that dating evidence is rather sparse; the general alignment with the nearby road and a coin from the base of the channel seem to imply a first- or second-century date (ibid.: 55). If the feature was contemporary with the road, it must have possessed some covering, maybe in the form of a culvert. Without substantial evidence for such attributes, some have seen this channel as potentially just a roadside drain.

However, beyond the suitable location there are other reasons to believe this is the primary aqueduct of the town. The profile of the conduit seems to match up with the discoveries outside the town, implying a continuity of scale that perhaps would not befit a simple drain. Moreover, the flint/mortar construction would have made it efficient at water retention and distribution. Rogers (2013: 83–86) thoroughly documents the extensive drainage programme that occurred in the town; much of this activity necessarily took place in the flood-prone areas of the town, close to the Itchen. The positioning of the newly discovered feature on the high ground, close to the gate, means it was probably not involved in this activity; thus its large capacity is conspicuous, and likely confirms the interpretation as a primary water supply.

The final feature worth noting is the road with which the conduit was aligned. The North Gate of the Roman town is agreed to have overlaid the northeast entrance to the Iron Age fortified settlement of Oram's Arbour. The street in question has been interpreted as the main Roman road leading from this particular entrance point. As a consequence, it is no surprise that evidence has been found showing that it was preceded by an Iron Age holloway (Ford and Teague, 2011: 179), which would have been a primary route in the landscape before the Roman settlement. The presence of the aqueduct along this route conforms to the idea that such conduits could be used to highlight points of significance within a local mythic landscape. As a holloway in the higher ground, it is possible that the route had some resonance with water before the creation of an aqueduct; a high volume of rainfall could have occasionally transformed the sunken feature.

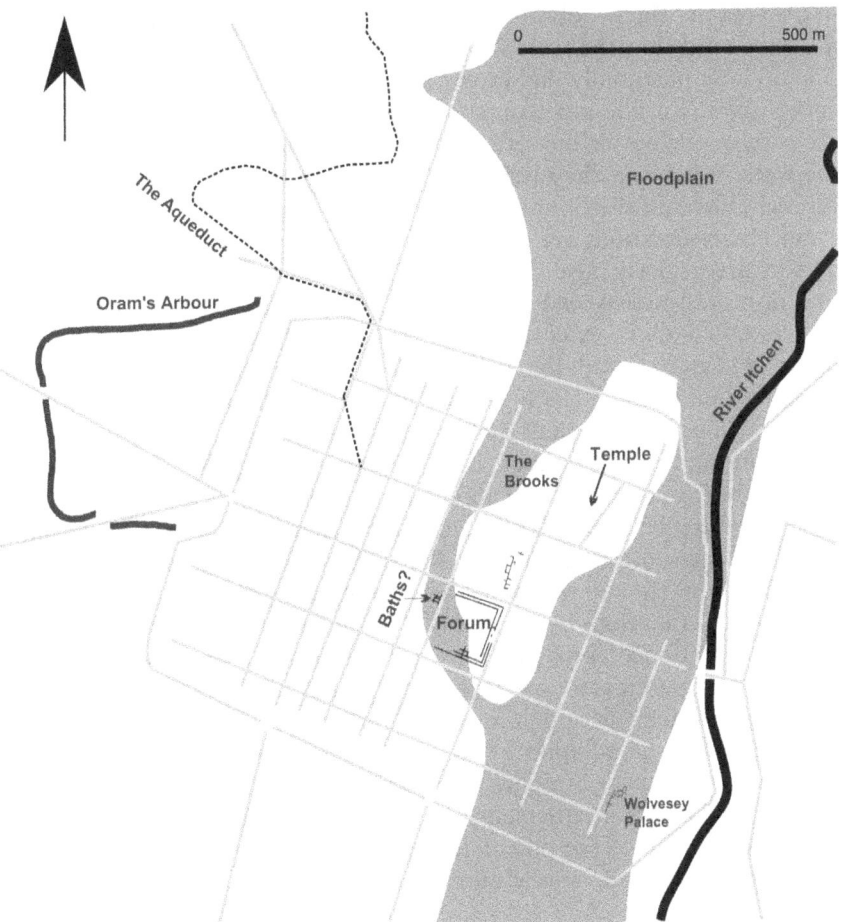

Figure 3.22 Plan of Roman Winchester with the potential internal route of the aqueduct proposed after recent excavations.

Source: Drawn by author after Ford and Teague (2011).

Taking all this into consideration, it is rather limiting to conceive this aqueduct just in terms of its structural grandeur. Stephens (1985a: 203) notes the 'enormous disparity' in both capacity and cost between this aqueduct and its larger contemporary supplying Paris (Lutetia). Yet the impact of this conduit from a more phenomenological perspective is surely just as marked as any comparable structure. Interestingly, there is evidence of water distribution further towards the centre of Roman Winchester. The remnants of wooden pipes were found in Lower Brook Street, running northeast to southwest, and a conduit with lead lining and an *opus signinum* covering was recovered south of Durngate (these finds are both in the area of 'The Brooks' in Figure 3.22) (Gunner, 1849). This is close to a temple, and thus the water could have been incorporated into religious practices. So potentially the water of the aqueduct was spread quite widely across the town: the pipes of Lower Brook Street indicate provision into the eastern reaches of the settlement, close to the river. If we conceive of the water taking on the meaning of the prehistoric landscape it travelled through, then this distribution is important symbolically.

Such considerations are only amplified by the fact that Winchester was already precariously positioned on the floodplain of the Itchen, so the addition of more water would not necessarily have been practically sound. In fact, as noted above extensive work was undertaken in the Roman period to drain areas around the Itchen, providing more stable ground upon which to build (Zant, 1993: 52). In the excavations of the Brooks area in 1987–1988 a monumental timber-lined conduit was found. In its first phase, during the first century, it was 1 metre wide and 0.55 metres deep, before silting and subsequent modifications reduced its capacity (Rogers, 2013: 85–86). There was also considerable dumping of material to make the landscape more viable (Zant, 1993). However, even with this investment of labour the site was never successfully drained, with a full return to waterlogged conditions by the end of the Roman period. As with a number of towns mentioned above, we have a situation where the manipulation of water forms a primary aspect of the activities taking place in the town during the Roman period. Moreover, yet again this is an example of a location where the addition of an aqueduct may have made construction even more difficult – bringing *more* water into a location already inundated. However, this settlement is not far from Chichester, and therefore could share some of the potential symbolic or religious connections with water proposed in previous sections.

Canterbury (Durovernum)

The Stour and the prehistoric landscape of Canterbury

Wacher (1995: 187) is among those who cite Durovernum as the first *civitas* capital in Britannia. With this assertion comes a host of problematic inferences that have characterised the understanding of the area. Foremost among these is a sense that the southeast of Britain was more receptive to an incoming Roman 'identity package' than other areas of the country. The writings of Caesar have helped to crystallise this way of thinking, with his observations regarding the similarities between those in the southeast and their Gallic contemporaries across the Channel, as opposed to the uncouth populace found further inland (Caesar, *Bellum Gallicum*: 5.14). Of course, Caesar's minimal forays outside the south of the country do not give us a sense of reassurance about the accuracy of such statements. In many ways these observations are representative of possible bias in the Roman worldview; the further north one gets, the more alien life will become. To an extent, the core–periphery models of the last 30 years promote this idea of a more continental, more Romanised, southeast region. It stands to reason that the Canterbury area had increased connectivity to the continent. However, if anything this should heighten the complexity of settlement within the Stour Valley, rather than making it seem more familiar or more 'Roman'.

Traditional discourse has interpreted Canterbury's relationship to the Stour as almost entirely practical. The river was navigable up to Roman-period settlement; thus in addition to easier road access, the valley-bottom site was a logical location for a regional centre. Indeed, the role of Canterbury as a key strategic point leading to the principal military supply base at Richborough is a primary part of the interpretation of both Wacher (1995: 191) and Frere (1967: 22). But while military motives seem to proliferate in historical accounts of Roman Kent, the story of Canterbury's development appears to be characterised more by local factors.

The recent discovery of a large Middle Iron Age settlement on the site of Turing College at the University of Kent (on the high ground directly overlooking the Stour Valley and modern city of Canterbury) adds to our picture of Iron Age occupation of this landscape. Considered alongside the nearby hill-fort sites of the later Iron Age at Bigbury and Homestall Wood, there is sustained occupation of the high ground above the eventual site of Roman Durovernum. Moreover, all these sites had conspicuous relationships with water. At the Turing College site, the Canterbury Archaeological Trust excavations discovered a notable perched water table that led to numerous springs and even waterlogging of certain areas (Lane, 2018). At Bigbury, recent research by Andy Bates (2018) has increased our knowledge of a complex series of earthworks (similar to those of other towns mentioned in this chapter) that engaged with the River Stour. And at Homestall Wood, a comparatively poorly understood site, lidar analysis has highlighted a clear spring and watercourse that runs from the centre of a large enclosure (Sparey-Green, 2018). While the nature of the relationship these sites had to water requires further investigation, it should make us question the connections Durovernum had to its waterscape in the river valley. Certainly, the site of Canterbury coincides with a rather dramatic braiding of the river and the creation of a large island (in addition to smaller eyots) that remains today (Figure 3.23). These are the types of conditions that are emphasised in this chapter as being fundamentally significant for people in the Iron Age. Rogers (2013: 52–57) summarises how past interpretations of the Stour in the Roman period sought to marginalise its presence. There has been a continued effort to suggest that it did not run directly through the town (see Wacher 1995: 429) and instead would

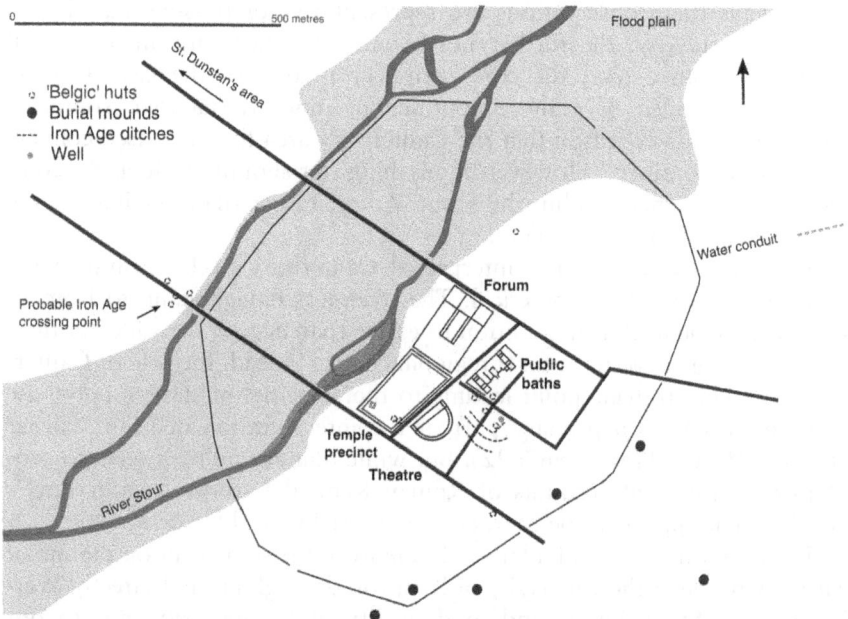

Figure 3.23 Plan of Roman Canterbury with significant features and the flood-plain of the Stour.

Source: Drawn by author after image provided by Canterbury Archaeological Trust.

have been diverted in a more practical way. However, the continuing archaeological work in the city has confirmed it is far more likely that the central areas of Roman Durovernum were characterised by a watery and marshy landscape (Rogers, 2013: 56).

There appears to have been significant occupation of this landscape, and some scholars have suggested this consisted of an organised *oppidum* (Wacher, 1995: 190). However, this interpretation may owe something to the consensus around viewing prehistoric settlements as proto-urban stages of later Roman towns (Haselgrove, 1989: 10). One thing that is easier to discern is the central place of the braiding Stour within the narrative of the Iron Age and the Roman occupation of the area. One of the crossing points of the river, close to modern Whitehall Road (at the London Gate), reveals evidence of a number of buildings, pits, coins and cremation burials (Weekes and Seary, 2011: 6). There is little substantive evidence for a ritual meaning of this crossing point, but it seems logical considering the fundamental importance it held in traversing the valley landscape. Conditions were right here to create a crossing, but such attributes may well have led to veneration in the past. The ongoing analysis of the Westgate

Garden area by Canterbury Archaeological Trust may further such interpretations.

Across the other side of the Stour, underlying the central areas of the Roman town, evidence has been found of a notable triple-ditched enclosure with associated roundhouses (Blockley *et al.*, 1995: 37). Rather than serving a defensive function, these ditch features have been interpreted as regulating movement within a scattered settlement (ibid.). Some coin moulds were found during the excavation of this site, suggesting either a mint or sophisticated metalworking in the vicinity (ibid.: 9). Moreover, there is a suggestion that the Roman temple precinct close to the bank of the river was preceded by an Iron Age religious feature of some sort (Wacher, 1995: 193). Movement between these two areas of occupation, on the east and west banks of the Stour, could have been a key feature of settlement. The intervening space between them was the marshy, flood-prone, island zone formed by the braiding river. These attributes may have encouraged the use of the place as a meeting point. Coming from either of the high-ground settlements, people probably had to move through the braiding zone before reaching any religious focal point on the east side of the Stour. Movement across the waterscape may have been a key feature of ceremonial processions, as Creighton (2006: 124–130) suggests in regard to Verulamium.

This liminal landscape is entirely incorporated into the Roman town, despite notable practical drawbacks. The western and southwestern parts of the town, for example, were particularly subject to flood events, becoming uninhabitable during the third century (Rogers, 2008: 47). Excavations at St Mildred's Tannery have uncovered a number of Roman-era buildings in this flood-prone area, but very little in the way of definitive structure types. One of the more intriguing of these is a large aisled building located close to the eastern ford of the river (Pratt, 2009: 231). Its function is hard to discern, but it was sizeable and appears to show signs of continued use when nearby buildings were being abandoned due to flooding (ibid.: 232). While it does not necessarily fit the profile of a temple, it is close to the river and the central religious area of the town. There is a possibility that this structure could be linked to the conception of the local waterscape. On the western crossing, close to the Iron Age activity mentioned above, a small bathhouse has been discovered; its meaning is discussed in a later chapter, but we should be wary of separating it from the local waterscape.

These points notwithstanding, the Roman town appears to embrace much of the same logic as the Iron Age settlement. The main road to London is postulated to have maintained alignment with the previous local trackway, exiting via the London Gate. In addition to the main road (from the Westgate), the settlement was characterised by the act of moving through the Stour's influence. The monumental features of the Roman town (the forum, temple precinct, public bathhouse and theatre) are built in a cluster on the eastern side of the river. In many ways their presence amplifies the importance of crossing water, and how that act is pivotal to the overall experience of Canterbury. The maintenance of fords may also have meant the visual spectacle of wet footprints processing into these central areas, further underlining the presence of nearby water.

The discovery of the Iron Age settlement on the university grounds may add to this interpretation. If one sees this newly discovered area of occupation as intrinsically linked to the lowland site at Canterbury, then movement between the sites would have been through modern St Dunstan's; excavation over a number of years has revealed extensive Roman-period burials in this area (Weekes, 2011). While such activity is common outside Roman towns, this side of the town appears to have been far more popular than the other external points of the settlement. Part of this is inevitably the importance of the road to London, but there is also a possibility that this continued activity preserves the importance of the route to higher ground. The ancestral importance of this route could have made it a particularly suitable place to bury the dead. This interpretation would further entrench the Iron Age sense of space within the experience of Roman Canterbury.

Water supply in Canterbury

For a location so defined by the flow of a river, evidence of water supply at Canterbury (Durovernum) is less prominent than at other sites. Excavations in the Marlowe Theatre carpark revealed evidence of Late Iron Age occupation. This consisted of a triple-ditched enclosure surrounding two roundhouses (Blockley *et al.*, 1995: 27); among these features were a collection of pits and at least one firmly identified well (ibid.: 41). A group of tinned-bronze horse harness trappings was found in the upper fill of one of these pits (ibid.: 44). This could of course be of ritual significance, but it pales in comparison to the rich well deposits noted at Silchester and London. Of course, the adherence of medieval and modern Canterbury to the same site as the Iron Age and Roman settlement means much evidence may have been lost, or indeed has not been accessible due to protected buildings and a lack of development-led archaeological investigation.

Concerning a potential aqueduct, Wacher (1995: 196) always saw the east or southeast area as potentially the most likely source of external water, and the sparse remains of water conduits found up to that time tended to support his proposition. There was evidence at the Riding Gate of timber pipes, and these seemed to link up well with the central bath area of the town (ibid.). However, in 1995 a reasonably substantial conduit was discovered leading from the northeast. It was found during excavations at Christ Church College, and consisted of a bed of gravel-filled lime-rich concrete capped by an *opus signinum* channel 0.16 metres wide and 0.15

metres deep. The channel was surmounted by a rough barrel vault of mortar-bonded reused Roman brick and tile. No dating evidence was found, but the method of construction and the material used suggested the conduit was in use from at least the second century AD (Jarman, 1997). The nearby extant conduit house that supplied water to Christ Church Abbey is a vivid illustration of the potential in this area (Bennett, 1987, 1990): it seems likely that the monks of the twelfth century continued the Roman tradition of water procurement in the area, as is often seen. Perhaps this could be a reason why more Roman evidence for an aqueduct has not been found: it may have been obscured by the later activity in this part of the city (including the cathedral). Nevertheless, it may also imply a perceived sanctity of the water in question, which was later incorporated into the ritual framework of the abbey.

Within the actual town, the paucity of evidence means that (outside the bathhouses) it is hard to get an impression of the flow and use of water in the Roman town. However, excavations at 3 Beer Cart Lane did uncover three water-tank type features with associated pipes (Bennett, 1980). These have been interpreted as fountains, and were located in the central temple precinct. In addition they are close to the theatre, perhaps being part of a monumental approach to this highly significant and dramatically large building (Tatton-Brown, 1977). It has been noted how the temple precinct district of the town was of Iron Age significance, and certainly had links to the nearby water of the Stour. The concentration of piped water from the surrounding territory in this area of Canterbury could have strengthened such associations. The theatre itself is of course seen as a sophisticated example of the Roman presence within the town. The potential for water being used in this arena, as with the example of Veru-lamium, could have changed the way people identified with and rational-ised the structure.

Bathing in Canterbury

The main bathhouse discovered at Canterbury was constructed at the start of the second century opposite the forum and close to the temple precinct and theatre (Blockley *et al.*, 1995: 84). In terms of scale, it probably com-pared favourably to the more celebrated examples of the Huggin Hill and Jewry Wall baths; the interpretation of the layout seems to fall somewhere between the two. The one unusual feature of the structure is the open *piscina* area, which is in a central courtyard. This has very few compara-bles in northern Europe, for the obvious reason of inclement weather. Little has been written about the reasoning for such an addition – indeed, most imply that it was simply not thought through. Undoubtedly this is usually because such features are characterised by a brief period of use before being replaced by a more suitable room.[11] While the Canterbury *piscina* is eventually a victim of refurbishment, this was after it had existed for some 200 years (ibid.: 96). This longevity makes it hard to see the feature as a miscalculation, but equally it seems unlikely that it would have been worth the cost of upkeep if it only served as a swimming pool (rarely frequented in the winter months). Bearing in mind the possibility of this water coming from special springs to the northeast of the town and the links the nearby temple precinct had to water, this pool could have been integral to other, more ephemeral, activities.

Water drained from the *piscina* out of its southwest corner before being channelled northwest (ibid.: 92). This is in the direction of the Stour, meaning the water would likely have passed through the forum/temple district of the town – something that is not without a symbolic resonance.

In line with this, Creighton (2006: 146) refers to the fact that, as with Silchester and St Albans, Canterbury has a significant temple enclosure with prehistoric roots that could have had an impact on the layout of buildings within the town. His account shows that Canterbury differs from these other two settlements in that the temple enclosure is in the centre of the town; although the presence of barrows on the outskirts of town is perhaps a local example of prehistoric ritual focus away from the central areas (ibid.). However, while the primary temple enclosure may be central at Canterbury, there is still evident proximity to a watercourse and a bathhouse, as noted at Silchester and St Albans. The link to these neighbouring features would have been heightened by the lack of surrounding walls around the bath *piscina* itself. It was seemingly open to the elements and, at least in the early periods, may have only been joined to the bathing complex at its northwestern corner (Blockley *et al.*, 1995: 90). It is possible that this gave the *piscina* increased visibility in the area, rather than being surrounded by the high walls of the bathhouse.

It must be emphasised that there was later evidence of metalworking in both the *piscina* and *laconicum* areas; this echoes the examples at London and Wroxeter. As mentioned before, it is possible that such activity could have represented more than opportunism. The later history of the bathhouse is somewhat hard to decipher, but it was apparently still standing well into the fifth century. Another element to consider is that eventually St Margaret's Church was built directly over the *piscina* area of the bathhouse. The church was constructed in the twelfth century but, as with many such buildings, it could have been a product of activities taking place in the area previously.

Canterbury is also unusual in that it has a number of smaller bath suites that are assumed to be private establishments. There is evidence of such features at St George's Street (Wright, 1948: 96), St Ragimund Street (Frere, 1947), Butchery Lane (Williams and Frere, 1949), the Marlowe Theatre carpark (Wright, 1958: 149) and St Mildred's Tannery (Pratt, 2009). Of these, the St George's Street example was quite large, with up to nine rooms incorporated in the complex. It was probably constructed during the second century and, as with the public baths at St Margaret's Street, continued well into Late Antiquity. The structure certainly had a significant overhaul during the fourth century, with the addition of a new cold bath, remodelling of the hypocaust system and a new entrance (Wright, 1948: 97). Of course, just because it is joined to what we assume to have been a private house does not mean it was off limits for people of the town. Indeed, the proliferation of smaller baths could be evidence of how in Canterbury it was particularly relevant to associate with these water-focused structures. The number of bathhouses in the town is certainly high when compared with other sites that have had far more excavations, such as Silchester (Hanson, 1971: 96). In the past, people have been tempted to explain this as differences in the extent to which the town was Romanised; the location of Canterbury in the southeast has traditionally helped to underline this argument of increased similarity to the rest of the Empire. Yet the settlement location was crystallised during the Iron Age and is deeply invested in the local waterscape, as discussed above. It seems equally possible that the compulsion to build bathhouses could have stemmed from local meaning found in the manipulation and presentation of water.

Cirencester (Corinium)

Siting of Corinium

Corinium's location in relation to the River Churn, the Daglingworth Brook and the surrounding hills is distinctly impractical (see Rogers, 2013). Reece (2003: 277) notes, intriguingly, how the town is located at a bottleneck directly in the path of floodwaters coming down the valleys. This rather illogical site choice is further undermined by the lack of suitable building ground. The pre-Roman site consisted of a narrow island with a gravel spine bounded by the two watercourses, which met to form a marshy swamp at Watermoor (Reece and Broxton, 2011: 5). Extensive reclamation would have been required. It is suggested that the Churn and Daglingworth Brook would both have been canalised and separated to flow around the town before meeting southwest of the settlement (ibid.). Any defensive advantage would undoubtedly have been outweighed by the continuous effort required to stem floodwaters, especially during the winter months.

In conjunction with the above, Reece (2003) highlights how Ermine Street takes two notable deviations on its immediate approach to Cirencester. The first of these (when the road crosses the River Thames) could be explained as practical: a suitable crossing point was selected. After this the road returns to its original projected course, which intercepted Fosse Way at a seemingly ideal site to the northeast of later Corinium. However, around 8 kilometres from the town the road deviates again into the flood-prone area where the town was eventually built. The other primary road approaches from the southwest and curves in a highly irregular manner before entering the town. As the whole island area was gridded, with no diversions, it has been suggested that the sole reason for the road taking this path was if the straight route were blocked on exit (ibid.: 278). The only other notable landscape features that could be in the path of the best-fit courses of Ermine Street and Fosse Way are the Tar Barrows.

Creighton (2006) acknowledges the possibility that avoidance of the Tar Barrows was an expression of respect, potentially legitimising the new settlement by linking it to ancestral symbols of power. Taking this further, while it is logical that the Tar Barrows could have played a pivotal role in the positioning of Cirencester, they may be just one part of a wider 'special' landscape. With all that we have considered above, the siting of the town on this island could conceivably be just as potent symbolically as respect for the nearby burials. The recovery of a relief depicting a reclining river deity is further evidence of mixing of traditions. Toynbee (1962, Plate 37, No. 31) describes this particular sculpture as stemming 'from a classical source, but the spirit and expression reveal a Celtic character'. This is one example of how the Roman symbolism of water is adopted and adapted by the northern European hand, combining the associations inherent in both traditions.

The deviations of the roads can also be understood in their relationship to the River Churn and Daglingworth Brook. The deviation of Ermine Street from the southeast meant that it required potentially two bridges/fords to enter the city. The second of these would have been located at the presumed confluence close to Silchester Gate. The projected bridge crossing would have been directly over this fusion point between these two bodies of water; something that is hard to ignore when the road has taken such a definite change of course. In addition, as Ermine Street exited Cirencester to the north it was bounded by the two rivers, so maintaining a direct relationship with water. Finally, the Late Iron Age *oppida* complex at Bagendon lies in this trajectory, possibly enhancing the meaning and experience of such a journey.

Water supply in Corinium

Neil Holbrook notes that small domestic wells are frequent finds in Cirencester (Holbrook and Salvatore, 1998: 25). Insula V of Corinium has drawn many parallels to Insula XIV at Verulamium (Holbrook, 1998: 209) due to the nature of the structures, which have been interpreted as a row of shops/workshops. Yet the similarity stretches further, in that Insula V at Cirencester also has a notable well. Located in Shop 3, it was initially capped by a monolithic well-head, the height of which was subsequently raised by a further two courses of stones (ibid.: 201). The objects recovered from within this feature were four pieces of lead pipe, three *tegulae mammatae* and the skull of a horse (ibid.: 204). The lead pipe suggests a broader water infrastructure in this area and the potential that this source could have been systematically tapped to reach a wider clientele. The horse skull echoes some of the pit deposits found in Verulamium, Silchester and Dorchester.

Other wells worth analysing were found during excavations by Cotswold Archaeology between December 2002 and January 2003, at the meeting point of Insulae IV and VII (Brett and Watts, 2008). An area was discovered that was interpreted as a garden; the excavators reached this conclusion through analysis of soil deposits and the form of nearby buildings (ibid.: 73). Within this garden area, two wells were discovered dating to the same period (late first to late second century AD). One of these produced a number of oxidised flagons, some of which were substantially complete (ibid.: 74). The excavators put this down to accidental loss, but their appearance echoes some of the examples from other towns; black burnished ware was also present in relative abundance, similar to deposits found in Dorchester. Added to this, a pit was truncated by one of the wells and contained scorched red clay indicative of *in situ* burning. This hints at other processes going on around the wells, possibly very practical (such as cooking or manufacture), but also possibly evidence of a greater significance for these features in the local community.

There is less definitive evidence of a piped water supply at Cirencester than at many of the other towns mentioned in this chapter. The only water conduits found in Corinium were associated with the Verulamium Gate. The canalisation of Daglingworth Brook and the Churn must have been necessary to create a sustainable settlement in the location. Heavy use of these rivers in the medieval period again suggests we are dealing with flows of water that are hard to see as purely natural. The evidence of pipes at the Verulamium Gate adds to the difficulty in effectively separating a 'water supply' from the river (Wacher, 1963: 19). It has been suggested that immediately outside the gate the Churn could have provided the source of water for these pipes. This effectively means that any special importance afforded to this river is inseparable from the water supply. Another alternative could have been utilising springs to the northwest of the town and channelling the water to a *castellum aquae* at the gate. Wacher (1995: 308) even suggests that the gate tower itself could have doubled as this water feature. Springs around Stratton could also have been brought to the town via the Gloucester Gate. All these interpretations would add to the deep relationship this town had to its surrounding water.

York (Eboracum)

Water and the wider context of Eboracum

Evidence of Iron Age activity in the area that became Eboracum has proved elusive, and it has been hard to get any picture of the transitional phases of occupation that can be seen in other prominent towns of the province. But to the east of the town, at Heslington East, ongoing excavations have unearthed a series of noteworthy Iron Age discoveries. This site is inextricably linked to water, with many springheads located in the area explored by archaeologists. Unsurprisingly, a number of wells and sustained 'waterholes' have been unearthed. The most significant of these seems to have been in use from the Bronze Age through to the Roman period (Ottaway, 2010: 20). Some important, seemingly cultic, finds were discovered in association with this feature; the most striking was a human skull, which still contained an impressively preserved brain (ibid.: 11). A Late Bronze Age hollowed-out wooden cylinder and a fourth-century Roman coin hoard have been interpreted as showing sustained ritual-type activity concerning the waterhole (ibid.: 20–21). A further small Hadrianic coin hoard was found close to another springhead (ibid.: 21).

In fact, as excavations continue a multi-period association with water features is becoming apparent. The Kimberlow Hill area of the site appears to have been a particularly fruitful source of underground water, with watercourses probably running down the slope to form a pool. A Roman stone-lined well was found on the high ground overlooking this zone and there is copious evidence of Bronze Age water features, ranging from unlined pits to more formalised wattle-lined wells (Chapman *et al.*, 2012: 298). What can be ascertained from these excavations is a distinct tradition involving well construction within the environs of the subsequent Roman town. The continued respect that appears to have been afforded to principally the waterhole feature is particularly relevant when we consider the conception of wells in nearby Eboracum. Also, it is possible that similar evidence has been obscured or destroyed by the substantial post-Roman activity and accumulated layers of York.

The potential significance of such water features in this area of Britain is strengthened by evidence from nearby Shiptonthorpe. This Roman-period settlement was built around the principal road that connected York and Brough-on-the-Humber. Some of the structures discovered by recent excavations appear to show an entanglement of local and imported building techniques, combining the British roundhouse with a more continental rectangular form. Millett (2006: 311) described this as a prime example of cultural hybridity, highlighting the orientation of these buildings as a deliberate marker of their mixed cultural genesis. The rectangular sections appear to be addressing the Roman road, whereas the circular aspects were seemingly more hidden from view from the primary thoroughfare. The most intriguing feature of Shiptonthorpe is an early fourth-century 'water-hole'. Despite its obvious practical functions, this pit appeared to be strongly associated with ritual activity (ibid.: 315), evidenced by the deposit of a number of complete ceramic vessels in the primary fill. The faunal assemblage was also striking, with a series of skulls (a bull, a stallion and two dogs) as well as a pair of cattle mandibles and a horse's right foreleg (ibid.: 314). After the waterhole was filled in the fourth century there continued to be an association with animal remains: there were an additional five burials, involving four pigs, a young cow and a calf, in the area where the waterhole had once existed (ibid.). While this is a much later feature than those found at Heslington, it shows that ritualistic connections to water supply features are present throughout the Roman period in the wider context of the *colonia*.

Water supply in York

While features of this ilk are not readily apparent in York, there are some hints that water possessed meaning in the town. One principal example is the dedication to Oceanus and Tethys (*RIB*: 663), the divine couple who spawned all the world's rivers and streams. Archaeologists have also found a substantial timber-lined well dating from the second century. It is located at modern Skeldergate, close to the bank of the Ouse (in the *colonia* area of the town), and is remarkably preserved. The well was some 6 metres deep and produced a series of well-preserved finds (Ottaway, 1993: 90); the most relevant for this discussion is a collection of around 20 shoes in the lower fills (MacGregor, 1978: 61). Such deposition may appear unremarkable, but it has been linked to a pattern of votive offering in temperate Europe and beyond. Indeed, van Driel-Murray (1999) postulates that the liminal nature of the foot, as an extremity of the body, is something that perhaps lends itself to identification with watery contexts like wells. Moreover, she outlines how the deposition of shoes could increasingly (during the Roman period) become a substitute for the human sacrifices of prehistory (ibid.: 138). Considering that the skull from Heslington East has been interpreted in sacrificial terms, it is possible that the Skeldergate well could still be fundamentally linked to Iron Age rites despite its later construction.

As with so many of these settlements, what seems clear is that an external water supply was present in York: definitive evidence of an aqueduct has not been forthcoming, but lead pipes have been recovered close to the bathhouse and terracotta pipes on the fortress side of the river. The latter were very similar to examples from the impressive aqueduct of Lincoln (Wacher, 1995: 175). A stone street fountain from Bishophill Junior is probably the best-preserved example of such a feature in the country

(Ottaway, 1993: 90); the fountain is similar in construction to examples in Pompeii and Herculaneum. The similarity is not without interest, as some of the fountains on these sites may have had ideological meaning inherent in their placement (Hartnett, 2008).

One of the more intriguing finds relating to the water supply is the large lead pipe recovered from excavations in Wellington Row. This mid-second-century conduit would probably have crossed the bridge over the Ouse and connected the fortress and *colonia* water supply (Frere *et al.*, 1990: 325; Wacher, 1995: 175). The powerful ideological potential of bridges was outlined most recently by Rogers (2013), and in this case a water conduit adds another dimension to the picture. We must at least consider the symbolic effects of joining presumably somewhat segregated communities. The increased hybridisation of the *colonia* and fortress settlements could have been expressed by this physical joining together.

York possessed an impressive sewer system that appears to have been constructed to drain the legionary baths (MacGregor, 1976: 1); one main channel of this sewer was 44 metres long with six tributary conduits joining it (Whitwell, 1976). The size of the system implies that a great deal of water was being used in the town. Added to this, examination of the silt deposits in the sewer revealed the seeds and pollen of plants which prefer limestone subsoil. Ottaway (1993: 31) notes that the source of any aqueduct must thus have been located in limestone country. This could have been in the region of Tadcaster, some 20 kilometres to the southwest.

The remains of the civic public baths at York are relatively sparse. It was unearthed during the construction of the Old Station in 1839 and again in 1939 when the area served as a bomb shelter (ibid.: 88). As a result, there is a lack of detail in the accounts and relatively little to compare to other examples. The one sizeable room we know of,has been interpreted as a *caldarium*. At 9 metres wide and 10.5 metres long, it is probably one of the grandest examples of such a room in the province (ibid.). Without a detailed floor plan it is hard to quantify what this meant for the overall structure; there is the possibility that this could be an example of local variation in the bathhouse. Intriguingly, the other known buildings in this area are temples that have been attributed to Mithras and Serapis. It is worth noting that excavations in the aforementioned sewer system did uncover a number of religious *intagli* and some fine gold pieces (MacGregor, 1976: 10). While these are generally interpreted as resulting from accidental loss, it is hard to give a definitive interpretation bearing in mind the fluvial action of the sewer. There may have been a ritual side to these deposits in the bathhouses of York, but we are unlikely ever to get a consistent picture of how many items ended up in the sewer.

Exeter (Isca Dumnoniorum)

Exeter (Isca Dumnoniorum) was another legionary fortress that became a prominent town in the end of the first and early second century. The legionary bathhouse was erected around AD 60–65 and a conduit was needed to supply the estimated 318 cubic metres of daily water demanded by the structure (Henderson, 1988: 98). The original conduit is thought to have been sourced from a copious spring in the area of the medieval St Ann's Chapel; again, this seems to be an example of later religious buildings utilising the logic of a water supply set out in the Roman period. However, when the first town was laid out in the early second century, a new aqueduct was constructed that approached the forum area from due north instead of entering via the *porta decumana* (Wacher, 1995: 338). This meant the aqueduct crossed the defensive ditch on a bridge supported by wooden piles before taking water into the western corner of the forum (Henderson, 1988: 115) (see Figure 3.24). This new water supply seems to have been sourced from springs further up the Longbrook Valley, in the vicinity of modern Well Street (ibid.). While the evidence for the course of the legionary conduit is rather sparse, this new direction would have required more extensive engineering for possibly little practical gain.

It is possible that a greater understanding of the prehistoric history of the site could help to explain this decision. Of course, the increasing influence of civilians after the legions moved away could have meant the adjustment of water flow to synchronise better with the surrounding landscape. Exeter lies astride a sloping spur which is thought to have carried a prehistoric ridgeway down to a ford across the River Exe (ibid.: 92). This suggests that the direction of the water flow would highlight an important route for people before the crystallisation of the Roman road system. The springs of the Longbrook Valley could have been important stopping points as people moved through this landscape. As a result, the direction of water supply could have considered the legacy of movement in the immediate area of the settlement.

In this regard, it is worth making reference to how the second-century aqueduct more thoroughly addressed Rougemont Hill. The existence of a later Norman castle on this high ground means there has been little modern excavation. Nevertheless, there have been some suggestions that it could have been a site of prehistoric activity. Within a 15 kilometre radius of Exeter at least eight Iron Age hillforts occupy similar positions overlooking the floodplains of the Exe and its tributaries (ibid.). Added to this, the hill is a source of unusual red trap stone (which is the reason for its name), which could have added to its significance in the pre-Roman period. Interestingly, this stone was used extensively in the construction of the legionary bathhouse, the only stone building in the early fortress (ibid.: 98–99).

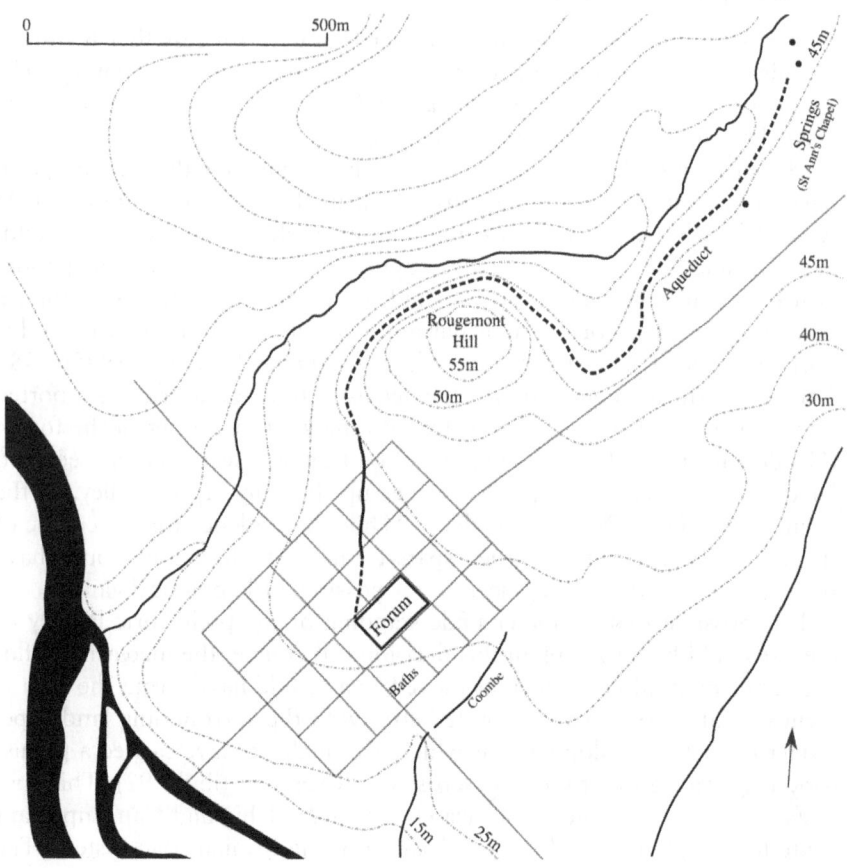

Figure 3.24 Course of the aqueduct into Roman Exeter.
Source: Drawn by author after Henderson (1988).

When Exeter eventually developed into a fully fledged civilian town, the bathhouse was altered to become a forum and basilica structure (ibid.: 112). The large outer walls were maintained, and the internal rooms altered to conform to a more orthodox forum plan. A new bathhouse was then constructed (probably in the late first century), and the aqueduct supply was rerouted to enter the town from the west to serve these central public buildings. The Exeter bathhouse possessed a *natatio* measuring 16.75 metres long and a little over a metre deep (Bidwell and Bailey, 1979: 122; Henderson, 1988: 113). This pool feature was surrounded by a pavement of sandstone slabs, which is thought to have carried a freestanding decorative colonnade and may have been open to the air. As with the *piscina* at Canterbury, we are quick to align such pools with classical bathing functions like swimming, but this standing water could also have had some symbolic significance. The date these baths went out of use is clouded with uncertainty, making it hard to propose any significant secondary activities in Late Antiquity (Rogers, 2008: 140).

Caerwent (Venta Silurum)

The native population of Caerwent is often referred to as being particularly resistant to Roman rule (Wacher, 1995: 378), but this does not appear to have stopped development of Roman-type structures. The copious supply of well water seemed to provide for most of the northern areas of the town. Evidence for an external piped water supply was originally derived from a line of iron collars belonging to a conduit outside the North Gate (Ashby et al., 1904: 93). This discovery was initially interpreted as a drain, but subsequent similar finds in a street north of the basilica (Ashby et al., 1909: 566–567) and on both sides of the north–south road to the southwest of the forum (Ashby et al., 1911: 426) confirmed an external piped supply. Unfortunately, evidence from inside the town gives only a loose impression of where the external flow was directed. A cement-lined conduit crossed the *palaestra* of the main bathhouse (Wacher, 1995: 382), and could have derived from one of the original conduits found to the north. The assumed *mansio* at the South Gate also seems to have received a significant water supply, with hypocaust systems and large drainage features nearby (Brewer, 2006: 15–16).

One of the more intriguing aspects of Caerwent is the evidence of the worship of native deities in the impressive temple structures. The Romano-Celtic temple, with trapezoidal enclosure, was probably not built until the early fourth century (Wacher, 1995: 386), but the discovery of an altar from AD 152 dedicated to Mars Lenus could indicate the deity worshipped in the later building. This deity is of interest, because one of the main references we have to it is at Trier, at a major sanctuary across the River Mosel in close association with a spring (Green, 1989: 64). It is interpreted as a healing deity, drawing on the original agricultural associations of Mars. As with Wroxeter, the linking of the town to a healing entity does raise the question of the conception of the local water, which is so often linked to such practices. If the water was being channelled in from a spring source in the nearby hills (Hanson, 1971), it is feasible that it could have been associated with the deity in question.

At Caerwent there are a number of wells that contained suspicious deposits; these features differ from many of the previous examples in their appearance close to houses. These are the sort of wells that one would usually interpret as being inherently practical; they served the daily water demands of the households in their proximity. Yet the list of objects discovered in them has much in common with those found in towns like

Silchester. Deposits from wells in the town include three fragments of a human skull, several ox skulls, numerous dog skulls, other bones, pieces of pottery, pewter jugs, a pewter plate, iron tools and fragments of a bucket (Ross, 1968: 262). In general these finds are in line with the consensus of Iron Age votive deposits, suggesting we may be detecting residences belonging to people local to Venta Silurum who maintained strong connections to their meaning-laden waterscape.

Other towns

Gloucester, Chelmsford, Aldborough, Brough-on-the-Humber, Carmarthen, Caistor-by-Norwich

At Gloucester (Glevum Colonia) an aqueduct is presumed by the evidence of iron collars from wooden pipes near the East Gate, plus several other examples within the town walls (Burgers, 2001: 98). A number of water tanks have been found within the town (ibid.; Wacher, 1995: 158); the most intriguing is a 9 metres wide example located in the fortress, possibly forming a terminal reservoir for an aqueduct (Stephens, 1985b). Wacher (1995: 159) notes that the water supplying this feature could have come from north of the town; he attributes this to a faint linear feature that has been detected running towards the east corner of the settlement from this trajectory. Hanson (1971: 367), on the other hand, suggests that the south of the town towards Robins Hill Wood would have been a more obvious source.

As at Canterbury, a later monastic water supply was located in this area and could have been based on a Roman precedent (ibid.: 59; Hurst, 1988: 66). This source was only a few kilometres distant from the villa at Great Witcombe, which was a site centred on a venerated spring. Indeed, there are a number of similar spring sites in Gloucestershire, such as at Chedworth Villa (Webster, 1983), Lydney temple complex (Ross, 1968: 22) and Lower Slaughter (Rattue, 1995: 29). It seems rather simplistic to think that water tapped from such sources would be devoid of meaning, thus the central tank in the fortress at Gloucester could have played a similar role to the examples in the Upper City at Lincoln – providing a focal point for water that had an enduring local meaning in the landscape. These comparisons perhaps start to break down the assumption that colonies were focused purely on creating an ideal 'Roman' settlement in a provincial setting.

The Roman settlement at Chelmsford (Caesaromagus) overlooked the confluence of the Rivers Can and Chelmer, thus controlling a potentially important symbolic point in the local waterscape (Drury, 1988: 1). Excavation at Caesaromagus has produced evidence of a number of votive shaft deposits. Outside the town a full horse burial was discovered in 1987

(Frere *et al.*, 1988: 458). Added to this, to the northwest of the *mansio* there was a disused well with a number of new shafts dug into it, housing the skeletons of at least seven foetal lambs, a horse skull and raven and cat bones (Drury, 1988: 9). These animals (particularly the horse) echo Iron Age tradition, even if Caesaromagus lacks a defined prehistoric precedent. This well was associated with a probable temple structure, so could have played some sort of public role in the life of the town. However, the comparative lack of archaeological evidence at Chelmsford makes further interpretation of water supply difficult.

Aldborough (Isurium) seems to have risen to prominence in the Hadrianic era. There is even less evidence of an external aqueduct here than at the primary town of the *civitas* (York); even the recent survey of the site led by Martin Millett has not revealed any sign of such a structure (Millett, 2017) However, a series of tanks have been discovered that were probably used to store water. It is unclear whether these were supplied by conduits, but they could have been important structures. The best-preserved example at the North Gate measures internally 2.7 metres by 1.8 metres and was made of stone (Wacher, 1995: 404). It has been suggested that the tanks could have been filled by run-off of water falling on to the main gate structures. While the evidence is too fragmentary to make any firm interpretation, such activity would be of interest considering the sanctity of gateways described elsewhere in this book. In later periods there is evidence of wells in Aldborough, so they may have been a feature of the Roman-period settlement (Ferraby, 2017). Nearby Brough-on-the-Humber (Petuaria) is another site that does not have particularly extensive archaeological remains for a likely *civitas* capital, but it was located on the corner of the Yorkshire Wolds beside the wetlands of the Humber. A mosaic at the nearby Brantingham Villa depicts eight river goddesses, in reference to the numerous tributaries of the huge catchment that drains into the Humber (Liversidge *et al.*, 1973).

There is a similarly scant picture on the fringes of the province at Carmarthen; the only evidence here is a leat that ran under the second-century walls, presumably representing an initial military water supply (Goodburn *et al.*, 1979: 272). But even here there was potential for water to be an important part of how people experienced the town: it was situated in the estuary of the River Tywi (James, 1993: 95) and thus at a confluence between fresh and salt water.

Finally, increasing archaeological investigation at Caistor-by-Norwich is helping piece together more evidence of what life was like in the capital of the Iceni. But little of this has thus far related to water supply/management. There were suggestions that a water channel stemming from the River Tas was cut outside the northern walls of the town (Davies, 2001, 2009: 171). This was refuted by Rogers (2013: 119), who suggests the aerial photography may instead show evidence of post-Roman activity.

Nevertheless, this is another example of a Roman town close to a confluence, at the point where the Tas and Yare meet (Wacher, 1995: 243), and thus a place already noted as having prehistoric ritual meaning. What is more, recent work has suggested that the River Tas likely ran close to the town in the Roman period, meaning it was a more prominent aspect of the urban centre (Rogers, 2013: 95).

Manipulating urban identities: multidimensional approaches to water supply

The study of urban water supply and management in Roman Britain has been primarily focused on how it represented a forward-looking subscription to an incoming identity package. Accounts of prominent water features like wells, aqueducts, sewers and bathhouses in the province have therefore principally been concerned with how they compared to similar features on the continent. In the author's view, outlined in Chapter 1, the frame of reference for this comparison has been structural scale and technological sophistication. This interpretative framework has resulted in British examples of water structures being largely divorced from the developments in their immediate context, and completely ignored in wider European discussions due to their lack of scale and sophistication. This chapter is a reaction to these approaches, highlighting the unique prehistoric context of what archaeologists have denoted as principal British towns, and outlining how their Roman-period water supply was often directly related to previous developments in the landscape. In these towns all the classic feature types like wells, sewers, bathhouses and aqueducts can be interpreted as having a unique contextual meaning dependent on their location. As a result, the ways they were understood within a community could have been profoundly different from the accepted portrayal of such features.

Wells

Obviously, there has been previous work on how Romano-British wells can be put forward as features that exhibit continuity with prehistoric practice. Fulford's (2001) discussion of the evidence at Silchester is a principal example of this, with the deposits of human remains and ambiguous holed pots in early wells proposed as an underlying Iron Age influence for the Roman town. Silchester stands alone in the sheer number of wells that have been discovered, and their prevalence in central insulae that span the Iron Age and Roman divide suggests connections between the urban waterscape of the town and its prehistoric precedent. The rich assemblage of metalwork finds recovered in the antiquarian excavations also suggest that these water supply features continued to have ritual prominence throughout the Roman period, countering an argument that there would be a 'progressive' movement towards solely practical uses for the wells. The wells of Southwark in London are probably the only genuinely comparable features that have been found in Britain. Southwark itself has been interpreted as a place that held pre-Roman meaning and had ritual focal points in the Roman period; it is also a waterlogged location that required sustained engagement in the form of bridging, revetments and canalisation. Thus it is easy to see how the discovery of wells with votive deposits can contribute to the discussion of complexities involved in the conception of Roman water supply. However, like Silchester, the rich metalwork deposits are all of a Late Roman date; so while they may be put forward as enduring symbols of ritual practices associated with water supply, they are not necessarily directly connected to the prehistoric context of the city.

The prevalence of wells in the Walbrook Valley should be of real interest to archaeologists trying to understand the nature of Roman London. There appears to be evidence of ritual activity centring on the Walbrook from prehistory and throughout the Roman period. Hence the nature of procuring water from this lowland area cannot necessarily be seen in mere exploitative terms. The wells at Queen Street, for instance, may be seen as a practical procurement zone, but they are also close to the religious area that would later house the London *mithraeum* (a religion with a noted link to underground water). The deep well at this site, with its human skull deposit, furthers the idea that this could have been a place marked with water significance. The pivotal role of the Walbrook should also make us question whether we have given enough consideration to the social side of the Gresham Street wells. Although their technological sophistication is worthy of analysis, they were a more substantial intrusion into the groundwater of this area than ever experienced before. If we assume that the Walbrook was meaning-laden in the prehistoric past, then the failure of these grand structures could have been explained by local folklore.

This chapter also documents wells in many other towns that appear to show either Iron Age ritual concern with these water features in proximity to later Roman water supply systems or an echo of such traditions at a later date. For instance, horse skulls, skeletons or trappings were found in wells at Chelmsford, Canterbury and Cirencester. These finds could be related to the worship of Epona, a 'Celtic' deity purported to have a prominent connection to worship of watery places. It is a practice seen at temple sites like Witham in Essex (King, 2005). Potentially having such practices and beliefs present in Roman towns gives us a sense of the diversity with which we should approach even poor examples of wells.

Similarly, the majority of towns covered in detail in this chapter have evidence of at least one well containing a significant collection of unused pottery, and a number of these also include ox or other animal bones.

These assemblages may be linked to ritual feasting traditions that became increasingly important during the Late Iron Age (noted earlier). However, perhaps more striking is the appearance of human skulls in wells in these towns. London and Silchester are mentioned above, but there are similar deposits at Caerwent, Heslington (in York), Dorchester and St Albans. Some of these might be the product of violent or clandestine activity, but that in itself is quite revealing of the associations people had with water features like wells. In the case of Dorchester and St Albans, it is likely to have been more overt organised deposition linked to the importance of the ritual enclosures interpreted at these sites. These Roman-period deposits echo the use of human remains in British Iron Age ritual practice, and thereby further solidify the idea that such elements of water supply were inherently linked to previous activities in the landscape.

While these deposits help us understand the legacy of interaction with groundwater at these sites, they are not necessary to demarcate importance. St Paul's Well in the Upper City of Lincoln was likely of prehistoric origin but became a clear nodal point of the waterscape in the Roman town, incorporated into the *principia* of the fortress and also the later forum. This prominence in the Roman period hints at a central role in the Iron Age. This would not be surprising considering the likely ritual activity taking place in Brayford Pool area, which is overlooked by the high ground that accommodated the well. The broader context of the surrounding region is rich with definitive evidence of water worship from the Bronze Age through to the Late Iron Age. Divorcing this well from its contextual meaning, as was the case until the work of archaeologists such as Adam Rogers (2013) and Mick Jones (2003), is undeniably limiting.

Of course, wells are only one aspect of the urban waterscape at these sites and, as explained in the previous chapters, are seen as likely receptacles for indigenous beliefs. In contrast, the piped water of aqueducts and sewers has been completely divorced from interpretations based on local context. The introduction of piped water represents an increased complexity in the relationship between a settlement and water. It could allow meaning-laden water to permeate the fabric of the urban settlement, and thus may have tangible impacts on the conception of many different structures and spaces within a given town. Taking these ideas into account, one can no longer look at the implementation of such infrastructure as merely a harbinger of full-fledged Roman identity. The technological innovation may stem from the incoming Roman influence, but the water harboured the story of the local landscape from years before.

Aqueducts and piped water

One must consider the impact that re-evaluation of piped water has on the comparative status of Romano-British water supply in general. Moving away from defining these structures by scale and embracing their inter-actions with surrounding landscapes changes their value for archaeologists. Of course, the pragmatic observation of Stephens (1985a) denotes an accepted truth: that a more dynamic landscape requires more sophisticated engineering solutions. Still, one has to consider cultural preference in the expression of meaning-laden water. Although there was no practical need for massive bridging structures such as the Pont du Gard in southern France, people in Britain could also have been very receptive to different forms of transmission for their water. The popularity of leat-style aque-ducts may be a choice based on relative price and the volume of water required, but their river-like appearance in the landscape is thought-provoking. Kamash (2012) notes how the assimilation of water technology in the east of the Empire had similar concerns at heart, and some elements of incoming Roman innovation were shunned due to general cultural pref-erence. Taking this into account, it is somewhat short-sighted to proclaim the aqueducts of Gaul as having greater importance than those in Britain.

There are many examples in this chapter highlighting the distinct importance of aqueducts at a very local level. Lincoln is the most obvious of these, because recent archaeological work has led to an increased under-standing of the undoubted importance of water veneration in the area from prehistory and throughout the Roman period. The assumed route of the *colonia*'s aqueduct implies it was a particularly impressive undertaking, but its real value is defined by the unique local connections to water. It is highly unlikely that one could understand the reasoning behind this water supply feature and its role in the settlement without considering these associations. As mentioned above and mooted by Jones (2003), the radical proposal that this aqueduct may even have taken water from the Upper City and directed it out of the settlement towards an external site of importance shows how unreliable traditional approaches can be in inter-pretation Romano-British water supply. Even if we remain conservative about the direction of water flow, this aqueduct was still siphoning a large amount of water through monumental sewers to the Iron Age ritual site of Brayford Pool/River Witham. That was an undeniably strong gesture, and the motivation for it has to take into consideration the local context.

While Lincoln may be the most striking example, towns such as Dorchester, St Albans, Wroxeter, Winchester, Colchester and Exeter all had small aqueduct systems that appear to be intertwined with their local Iron Age contexts. In the cases of Dorchester, Winchester and Exeter there is direct interaction with previous settlement features. At the former the conduit is incorporated into the fortifications of Poundbury hillfort; at Winchester it is the way that the Roman aqueduct follows the line of a hol-loway that ran directly into the Oram's Arbour enclosure; and at Exeter the civilian aqueduct of the second century was rerouted to address the probable prehistoric focal point of Rougemont Hill. The example at

Dorchester, of course, has the added intrigue of this water then being focused on an area in the central town that appears to be ritually significant. St Albans has very apparent precedents for large-scale interactions with the local waterscape: the dyke systems appear to highlight connections between areas of water significance, and the activity centred on the Ver is overtly ritualistic. One of the probable sources for an external aqueduct was an upriver location with a notable temple complex. The internal water sources and drainage features also seem to play a part in emphasising the pre-existing relationship between St Michael's enclosure, the River Ver and Folly Lane. Similar practices can be proposed at Colchester, where the 'Claudian leat' ran from the influential peripheral Iron Age site at Sheepen to the symbolically profound Balkerne Gate. Moreover, the movement of water out of the town from the supposed *mithraeum* opposite the Temple of Claudius confirms distinctly non-practical behaviour regarding water in the settlement. Even at Wroxeter, where we have traditionally given little focus to the pre-Roman context of the settlement, there is a degree of overlap between Iron Age practices of symbolic control of water in the area and the incorporation of the aqueduct and Bell Brook into the central infrastructure of the town.

Other sites, such as London and Chichester, may not have had the clear external conduits of the towns above, but they appear to have harnessed their respective local rivers more extensively to create internal aqueducts of sorts. In such a process there is a very direct incorporation of the local waterscape into the fabric of a town. In the case of the Walbrook ritual activity is related to its waters from deep in prehistory, and this is unlikely to have become immediately irrelevant in the Roman period. Indeed, the clustering of religious structures close to the river, in addition to the potential votive deposition happening on its bridging points, suggests that this water supply could have changed the way that people experienced the buildings of central London. The proposed Roman-period redirection of the Lavant at Chichester may be of similar importance. As a chalk river it possessed unique properties, and had a connection to the South Downs that would have played a part in the local sense of place and time. The likelihood of the local ruler Cogidubnus having a role in the construction of the urban waterscape suggests this is about remaking the Iron Age context in a new era, rather than just an unrelated uptake of external fashion. There is also a possibility that the Lavant was part of wider processions through this landscape, which is notable for its water-based shrines like those on Hayling Island.

Bathhouses

Much water in Britain appears to have been directed towards the urban bathhouses. Chapter 2 outlines how bathhouses in the Mediterranean were structures that encompassed a diverse and hybrid array of functions. This adaptability was primarily motivated by the unique local concerns of a particular location. The pattern of benefaction elsewhere in the Empire suggests that wealthy individuals contributed significantly to the development of bath structures in their localities (see Mackie, 1990). Some say that Britain represents a very different example, with funding for such structures coming from military or civic sources rather than local elites (White, 1999: 289). Still, the extent to which one can separate public office from the idea of the local elite is somewhat debatable. Furthermore, epigraphic evidence relating to the construction or restoration of public bathhouses in Britain is comparatively poor (Blagg, 1990: 15). This lack of evidence of dedication may well be paradoxical proof of a stronger native/local influence on bathhouse construction. Mattingly (2008: 67) suggests that in Britain the epigraphic tradition was not necessarily as thoroughly accepted by the local elite. If this is so, the lack of definitive evidence for many bathhouses becomes less significant; it could just indicate that the individuals who created and cared for them did not see the need for inscriptions. Regardless of the particulars involved in interpreting the epigraphic evidence, it seems logical to believe that influential people of British descent played a role in 'civic improvement'. This acknowledgement must make us question the admittedly less comprehensive evidence we find in Britain. It opens the door to local agency being an integral part of the creation of these structures (DeLaine, 1999b: 11), and it is consequently reasonable to suggest that these buildings had some relationship to the legacy of belief within their immediate landscape context.

The examples of this chapter show that in many prominent towns in Roman Britain there is a link between bathhouses and areas of prehistoric landscape importance. Structures from Lincoln, St Albans, London, Silchester, Dorchester, Chichester and Canterbury are all discussed in these terms. Moreover, there is a case to suggest that in the Roman period the majority of these towns had religious processions of some form that incorporated their respective bathhouses. Of course, it is also proposed in this book that the flow of water though aqueducts could have carried meaning, and thus defined structures away from an original meaning-laden natural source. If this added layer of interpretation is considered, many of the bathhouses in towns like Wroxeter, Exeter, Winchester and York become highly relevant. With this in mind, it is hard to reconcile the uptake of bath buildings in Britain as primarily an act of cultural alignment with the Mediterranean; instead, it is plausible to think of them as intensification of activities concerning water in the various local landscapes of Britain.

That is not to say there was no Roman influence on the buildings, as there surely was: the layout of British bathhouses loosely complies with a Mediterranean standard. But it must be noted that we are sometimes guilty of imposing this accepted rationale on room types in bathhouses with very little positive evidence. The British evidence is often so sparse that we assume similarity with other bathhouses to gain some clarity on the purpose of each room. But by committing to this method, unusual additions or activities automatically become signs of 'decline' or some lack of comprehension of classical standards in the provincial periphery. The evidence of metalworking in the Huggin Hill baths of London, for example, is found alongside the possible continuation of a bathing function in other parts of the building (Rogers, 2008). It might be natural to think that this suggests a downsizing of the baths and opportunistic exploitation of the structure. However, if the building was never used in a strictly traditional sense, it is hard to be specific about such issues.

Health is another local aspect of water particularly linked to bathhouses. This has been ignored in Britain, mainly due to the spectre of the formidable evidence of Aquae Sulis (modern Bath). The richness of this site has somewhat hijacked the issue of health rituals in the civic baths of the province. Of course it has some grand instances of syncretism between native and Roman beliefs that surround water. Such examples are testament to the enduring strength of Iron Age belief systems, but they also set an unfair precedent. Aquae Sulis was a thermal bath, and evidence of these is rare in Britain, thus it does not bear a tremendous amount of relevance for the more 'standard' bathhouses sourced from fresh water in many of the towns mentioned. The occurrence of oculist's stamps (*collyrium* stamps) at the Wroxeter baths is perhaps the best example of these less visible practices taking place (Jackson, 1999: 110). It has been noted previously how the Wroxeter aqueduct has been interpreted as terminating in a temple structure that could have been related to eye complaints (Wacher, 1975: 442). This supposed temple was only just north of the bathhouse. Subsequently, it is plausible to suggest a connection that could have been integral to the bathing experience. At towns like St Albans, where we have bathing structures incorporated into the ritual waterscape of the Ver, or Lincoln where the water had a pronounced value far back into prehistory, these sources could have had transformative or healing values that are hard to discern from the archaeology. Despite the Roman derivation of the bathhouse structure, healing functions could be representative of local knowledge and associations that stem from the Iron Age, and they would surely have been a transformative force in the use of space in the building (as discussed in Chapter 2).

Essentially, this chapter presents the idea that meaning-laden interactions with water in prehistory have relevance in discussions of the subsequent establishment of Roman urban water supply and usage. Hitherto this has rarely been the case. However, that is not to say that the Roman interactions represent strict continuity with practices of the Iron Age. What we see in the case studies is an intensification and monumentalisation of water interactions, for perhaps many different reasons, but with a consideration of this local context remaining apparent. This is the hybridity mentioned in the opening chapter, describing a diversity of expression in the various towns of the province and combining motivations and associations of the diverse populations that were present in these settlements. The next chapter highlights and explains some of these differences, connecting these case studies to the incoming motivations of Roman imperial power discussed in Chapter 2. It suffices to say here that traditional approaches to British water supply, which have looked outwards to create unfavourable comparisons with continental Europe, may have grossly underestimated the value of these developments and the subsequent impact they had on the experience of urban space in the Roman period.

Notes

1 Another marble base has been recorded in Insula VIII. Niblett (2001: 77) suggests that Watling Street could have been lined with such fountains.
2 Wheeler and Wheeler (1936: 104) note several examples: ALBUS, ALBUCIUS, MAGIO, MICCIO, UXOPILLUS and possibly MARCELLUS.
3 Excavations at Heybridge (Atkinson and Preston, 1998), Oakridge (Oliver, 1992) and Chells (Going and Hunn, 1999) all uncovered wells with conspicuous proportions of samian.
4 A style of depiction practised in the Mediterranean context (Clarke, 1991: 312).
5 There have been challenges to this interpretation – most notably Knüsel and Carr (1995) – but ritual deposition of the prehistoric skulls in the Walbrook remains a strong possibility.
6 Cosa is the one other place where archaeologists have recovered these bucket-chains (Blair, 2002).
7 Interestingly the buildings at 1 Poultry and Pudding Lane are similar structurally (Hill and Rowsome, 2011: 372). It is possible that their shared circumstances, next to prominent river crossings, could have had an effect on their form and function.
8 Insulae XXI, XXII, XXIII, XXVII and XXXII are all mentioned (see Fulford, 2001).
9 Burial mounds continued to be a feature of this location into the Roman period (Warren, 1913).
10 Bearing in mind the Claudian associations of the town, it is worth mentioning how that particular emperor also established an aqueduct as a monumental entrance to Rome. The Porta Maggiore carried the Aqua Claudius and the Anio Novus into the city, and its symbolic importance is perhaps denoted by its prominence in the late Roman period.
11 A good example of this quick abandonment is the similar feature found at Gadebridge Park villa (Neal, 1974: 68–69).

Chapter 4

The value of water and new approaches to urban space

This book sets out to research a number of urban waterscapes in the towns of Roman Britain and illustrate their hybrid nature. The term 'hybridity' has been used to encapsulate the great degree of complexity that is inherent in the cultural associations and response to the use of water in these contexts, primarily during the first and second centuries AD. In this book, one of the primary criticisms levelled at past work is the tendency for writers to rely on a set of simplistic building archetypes when interpreting Roman towns in provincial contexts. These familiar structures have come to represent a set of universal values that are to be expected in every town, but which are often informed by outdated ideas of uptake/rejection/continuity of monolithic identities. It is suggested in the opening chapter that a realistic expectation of any settlement in this era is a layered and multi-dimensional experience that could represent a coming together of many cultural associations (from both local and incoming sources). This view of the urban experience is the core reason for using the term hybridity. The following discussion explores how urban waterscapes covered in the previous chapters can demonstrate a variability and complexity that has not been considered in past work. Moreover, the emphasis on divergent local approaches to water is discussed as a starting point for new ways in which the Roman town can be relevant in ongoing discussions about the future form of urbanisation.

Water and hybridity in the Mediterranean

As outlined throughout this book, there has been a tendency for archaeologists to favour practical reasoning when explaining the use of water in Roman towns. This has mostly been due to the way many writers still identify the Roman period with the modern world: because we emphasise practical reasoning for the supposedly analogous structures of today, similar motivations for their creation are expected to translate directly to the Roman era. Any consideration of symbolism has been skewed towards modern reasoning, with issues such as conspicuous consumption and largess (public munificence) often taking centre stage. Such an approach to water is not entirely wrong, but it represents a limited discussion of a multifaceted subject. Relying solely on these explanations essentially endorses a uniform caricature of Roman beliefs and motivations that is not easily identified, even in Rome itself. Moreover, it propagates a sense of distance between 'human-made' structures and their 'natural' surroundings. When authors emphasise practical reasoning behind water structures, they invariably cast the natural surroundings as something that needed to be overcome or dominated. This becomes a one-sided relationship that does not thoroughly explore the power and symbolism inherent in such waterscapes in local British prehistory, and similarly complex incoming Roman beliefs. There are practical reasons for all the water elements explored in this book, but acknowledging this does not rule out deep engagement with the symbolic power of the local waterscapes examined. Furthermore, this entanglement imparts a degree of individuality on these structures that takes their interpretation outside the realms of previous polarising labels of 'Roman' or 'native'.

Chapter 2 explores classical attitudes towards wells, aqueducts, bathhouses and associated sewers. The aim of this is to expose how the Roman view of these features was far removed from any simplistic definitions based on parallels to modern practicality and 'civilised' living, which are often used to justify such building projects in provincial settings like Britain. Similar characteristics were proposed for wells and aqueducts, two primary elements of water supply frequently associated with urban life throughout the Empire. It was argued that in Rome it was abundantly clear that underground water was imbued with significant associations, ranging from the medicinal to the miraculous. Twentieth-century scholars tended to work on the premise that such meaning was reserved for 'living' spring sources and thus was separated from human-made forms of water procurement (Wissowa, 1912). In this book the reality of such a divide is questioned. This is especially true when one considers the practical difference between a well and a spring that was being used by a community. Of course, it could be asserted that a well represents a structure that assists the extraction of water from an underground aquifer, while a spring source is water that has already reached the surface through natural processes. However, the best place for a well is somewhere that has a spring source close to the surface. Moreover, when one formalises a spring to be used by a community, its structural framework could reasonably be called a well. In fact, considering the extensive references to efficacious springs in Italy, it seems entirely possible that there were significant 'human-made' additions to their immediate surroundings. The result is that the lines between what was 'natural' and what was 'human-made' became blurred. Thus it would be unfair to separate the meaning of wells from the general portrayal of a powerful and highly localised underground waterscape. This is apparently confirmed by sources like Seneca (*Quaestiones naturales*: 3.11.4, 3.26.6), who notes such supernatural associations specifically with wells.

Applying such logic to an aqueduct and its extended piped water system is equally revealing. It has become customary to see these systems primarily as practical conveyors of large quantities of water to an urban setting. The symbolism attributed to these structures has mostly related to issues of scale and conspicuous consumption. Yet the fact remains that Roman aqueducts in Italy were often sourced directly from springs; the same springs that are portrayed as universally meaningful and in possession of varied powers and local associations. In the case of Rome there is very little to suggest that the creation of an aqueduct somehow lessened or eroded the power of the spring from which it was sourced; indeed, there is much evidence to the contrary. If its movement in a conduit did not adversely affect the perception of spring water, then the reasoning behind the creation of an aqueduct becomes a far more complex cultural decision, even in the heart of the Empire. It also means that it is insufficient to describe an aqueduct as a standardised object that has a consistent meaning in any setting. These features had human, natural, practical and symbolic associations that varied depending on the surroundings and the social milieu of the people interacting with them.

This reappraisal of the value attributed to Roman water supply networks was proposed as having a profound impact on the way we portray bathhouses. If such water maintained the importance and power of its source, then its incorporation into a bathhouse would have been significant. This is perhaps reflected in the way that the classical sources refer to aqueduct waters as possessing the purity of the springs from which they were sourced; almost as if the spring fount had been transported to Rome and was housed in the impressive *thermae*. On a larger scale, this represents an aspect of bathhouses that has been underemphasised in the archaeological dialogue. Many writers have portrayed them as buildings with consistent cultural and practical functions. However, the analysis of classical references to baths presented in Chapter 2 gives a sense that the meaning of these structures was highly malleable and often dependent on local factors. Therefore, just as wider political aims were easily expressed in their construction, the power of water in particular landscapes could also be effectively focused and respected in the uptake of such a structure.

Even in the heartland of Rome it would not be accurate to describe any of these features as entirely 'human-made' or 'natural', symbolic or practical. It is suggested here that this represents a degree of hybridity. However, this hybridity is not a consistent mix of values and associations, but instead dependent on the circumstances into which these features were established. Two aqueducts may have borne water that was equally important symbolically, but the specific importance of the water was determined by, for instance, the properties of its source and the individual journey it made before reaching a settlement. Such thinking inevitably changes the way we look at the implementation of wells, aqueducts, sewers and bathhouses in provincial settings. It means that even if one were to take a completely one-sided 'Roman' view of these structures in Britain, the analysis would have to be open to similar levels of integration with local contexts as seen in the Mediterranean. It thus becomes increasingly difficult to make sweeping generalisations about the universal motivations for the uptake of water features in a British context, because each example is entrenched in a different waterscape. On a more fundamental level, the receptive nature of these incoming Mediterranean associations means it is likely that similar British beliefs had an impact on the formation of Roman towns and their relationship to water.

Hybrid motivations and functions for water supply in Britain

The previous chapter outlines how the implementation of Roman water supply often interacted with areas of local waterscapes that communicated meaning in prehistory. These developments often appear to have heightened and amplified the relationship people had with water in these landscapes, bringing it further into the experience of settlement. Some Roman towns were built directly on, or beside, *oppida*-type settlements, which in the Iron Age possessed central areas where water was already an integral part of the matrix of habitation and land use (towns such as Canterbury, Leicester, Winchester, St Albans, Silchester and Chichester). Other sites had more ephemeral activity in the Iron Age (e.g. London, Lincoln, Wroxeter, Exeter and York), but in the Roman period became notable examples of integration between a settlement and the local waterscape. The overarching reasoning behind this could be just as layered as the direct experience of the features in question. Instead of emphasising Roman or native dominance, we are seeing a more hybrid picture of urban development.

Incoming motivations

As outlined in Chapter 2, the incoming Roman authorities at these sites were fundamentally receptive to and practised in the manipulation of water for symbolic purposes at the local level. Hence it is not a great stretch of the imagination to think that when creating new urban centres, such ideas could have been seen as an ideal way to promote legitimacy, and therefore secure a greater degree of stability for the new foundation. Authors like Rogers (2013) and Jones (2003) discuss the idea of imperial manipulation of ritually significant British water sites. However, this often emphasises the ideas of dominating and exploiting the landscape as a statement to a conquered population. It is thus tempting to explain development at a site like Lincoln as a confrontational and domineering gesture on behalf of this imperial authority. In this way of thinking, an important Corieltauvi site was captured and replaced by a distinctly Roman central place, to underline the new political hierarchy with its own symbolic priorities. Yet maybe in British towns we can also see more nuanced and considered actions of the Roman authorities: identifying a critical area of cultural proximity that could be used to aid more peaceful centralisation and organisation of territories. This would have meshed with their own religious and symbolic framework, but also promoted an element of legitimacy to satisfy the wider population and secure political and economic aims.

At Lincoln the development of the hybrid urban waterscape may reflect the willingness of the Roman authorities to interact with such local beliefs to establish, legitimise and bolster a tenuous new colony. The name Lindum, as an amalgam of both Latin and native language that directly references the watery focal point of Brayford Pool, might be an obvious marker to this hybrid characteristic (Rivet and Smith, 1979: 392–393). As outlined in Chapter 3, the incorporation of locally important water sources may well have turned the modern Upper City area into a place more akin to a shrine than a central administration space (Jones et al., 2003b: 7.22). The forum and the bathhouse probably possessed clear imperial imagery, but also appeared to celebrate the traditional importance of the local waterscape; the waters from St Pauls Well, the Blind Well and the aqueduct all took centre stage in these buildings. One could perhaps make a similar case for the possible temple precinct that is established in the Lower City; the construction of explicit water focal points – including a fountain and an additional bathhouse – may have furthered this incorporation of the local waterscape into the Roman-period urban framework. Even the development of the Wigford Causeway, while easy to interpret as an overt display of imperial power, can also be seen as part of a long sequence of bridging on the Witham, including the nearby Stamp End Causeway which was clearly of ritual significance in the Iron Age.

While Rogers (2013) rightly documents the significant reclamation and subtraction of the waterscape surrounding the Brayford Pool during the Roman period, the creation of the municipal water network actually put people in deeper contact with water than ever before at this site. The increasing interaction with water in this landscape during prehistory suggests that local people may well have embraced these new developments on their own terms. The probable continuation of religious activity in the Brayford Pool area after the construction of the Wigford Causeway is potentially an initial sign of such attitudes. The expansion of the settlement outside the area of the legionary fortress perhaps also suggests an attraction of local people to the site. Accordingly, it is very difficult to generalise the experience of Lincoln by relying on the imposition of a clear and dominant new Roman identity on a symbolically important site. It is tempting to view Roman colonies as one-sided cultural impositions on provincial landscapes. However, Lincoln might be an example of a hybrid urban landscape, with principal inspiration from the incoming Roman presence coupled with a keen awareness of how local traditions could create a successful settlement. It was a state of affairs that goes some way to limiting alienation, satisfying a highly superstitious Roman military from a religious perspective[1] and also providing a platform for economic, political, and religious stability among the local populace.

Ancestral landscapes

The Roman-led impact on waterscapes is just one angle on the subject. Britons who were local to these emerging urban centres would have had views and associations that affected the form and experience of any town in this era. When it comes to the motivation behind constructing the types of water structures explored in this book, the perspective of these people has been grossly oversimplified or ignored. The reality is that, as discussed above, because the Roman traditions involving features like wells, aqueducts and baths embraced a considerable degree of local variation, their purpose in Britain was undoubtedly influenced by the attitudes of local people. It is affirmed throughout this book that water appears to have been meaning-laden in many different areas of Britain for an extended period of prehistory, including the Iron Age. How people chose to engage with this special medium varied. One reason for this may be that the appearance of water in any landscape is subject to great changes depending on many different natural processes. As a result of such variability, how people interacted with water was also subject to change over extended periods of time.

It has been mentioned how Bronze Age interactions with water were often characterised by votive deposition of weaponry and potentially had a marked relationship to funerary rites; the appearance of burial mounds built near the water may be a sign of this link. However, in the Iron Age there appear to be developments in the way people viewed water, potentially expressed through the deposition of a wider selection of items, but also increased structural interactions with water. By the Late Iron Age the relationship to water seems to have become closer and more intimately involved in life and settlement. Bearing this in mind, the general character of Roman-period water interaction in Britain was not necessarily at odds with the changing relationship people had with water during the pre-Roman era. As explored above, the malleable functions or associations that were characteristic of these incoming urban water features could well have furthered such integration of local perspectives. There was undoubtedly an influx of new ideas, with bathhouses and aqueducts as probably the most prominent examples, but the way people experienced these spaces did not necessarily take inspiration solely from incoming practices.

There has been much debate on the way that Late Iron Age communities functioned. In some areas there appear to be recognisable top-down hierarchical structures emerging in the form of local chieftains. In the southeast of the country we witness the rise of powerful figures like Cogidubnus, who may well have been the product of such systems. Creighton (2000, 2006) endorses the view of this native elite being well versed in the manipulation of symbolic cultural features from their immediate local landscapes and also incoming Roman traditions. He casts these individuals as playing a pivotal role in producing legitimacy for many emerging urban centres, by combining incoming ideas with their ability to call on knowledge of how power was traditionally produced in their regional centres. The existence of such groups or individuals creates another layer of perspective when considering waterscapes. Often their role is designated as being the driving force behind the uptake of Roman lifestyle choices; these native elites wished to mimic their continental peers and aspire to the civilisation of Rome (see Millet, 1990). Some of them may have even grown up in the Mediterranean and wanted to 'civilise' their native lands along similar lines. These motivations may be compelling to a modern audience, but are also primarily an endorsement of Romanising dialogues that are unhelpful in documenting the changes of the period (see Webster, 2001). However, re-evaluating the value of urban water features along the lines of this book gives us ample room to suggest the motivation of such local elites. The cultivation of urban waterscapes presented an opportunity to project power on different levels. They would utilise incoming methods to signify relevance in the broader Empire, but the message would maintain relevance for the local population in a familiar language of landscape interaction derived from the Iron Age. Consequently, it could be seen as a way to centralise their power base to fit better into the Roman provincial organisation while not neglecting the enduring symbolic and ritual currency of waterscapes they had likely used effectively in the pre-Roman period.

It is suggested by Creighton (2006), for instance, that the development of Verulamium was based on the ritual processes enacted to venerate a tribal ancestor buried in the Folly Lane enclosure. The interpretation of this book is that the immediate waterscape was a tool in the production of this ancestral power, with the burial enclosure highlighting the central importance of the Ver, in addition to a combination of Roman and pre-Roman monumental works focusing the waterscape on to this processional route. The continued manipulation of this waterscape was a way for local people to forge deeper links with this foundational strength. The apparent use of Roman symbolism on coins minted by Catevellauni leaders in the

Late Iron Age shows there was a distinct awareness of continental trends in these areas of Britain. The possibility that such individuals even visited Rome would only aid such developments (Creighton, 2000). This means that the uptake of Roman techniques and building forms does not have to derive solely from the direct presence of a continental host. The hybridity of the Verulamium waterscape results from a combination of incoming knowledge and the established pattern of interacting with the local waterscape.

While Verulamium may be the most convincing example, similar motivations may have inspired developments at Dorchester, Canterbury, Chichester and Silchester, all of which have interesting relationships between central water features and proposed ritual activity during prehistory. In these towns the incoming structural forms and techniques may have allowed a more profound integration of water into the lived experience of the place. Again, these are still hybrid waterscapes, rather than continuations from previous eras. However, it is possible that the changes at these sites were inspired by local beliefs. This leaves us with a series of urban features that may have typical Roman form but were potentially being utilised to express long-lived parochial systems of power and legitimacy through the enduring medium of water.

Finding water and creating communities

Of course, it is not certain that everywhere in Iron Age Britain had a clear-cut and motivated hierarchy who sought to gain more personal advantage from the cultivation of symbolic places. Areas of the country that perhaps lacked a chieftain structure to dictate on a regional level may have consisted of many smaller communities characterised by less rigid hierarchical structures (Hill, 2011); kinship links or common interests perhaps bound these groups. This societal structure could have been reinforced and maintained by the coming together of disparate communities at particular points in the year, perhaps related to agricultural practice or established ritual events. The value of water in Iron Age society shows that it may have been a focus for large-scale activities that connected such communities. For example, it has been suggested that the Late Iron Age enclosure at Stanwick North Yorkshire may have developed around a sacred bog: the source of the Mary Wild Beck (Willis, 2013). There are a number of instances mentioned in this book where the coming together of people close to water may have been the initial catalyst for the creation of a settlement and monumental structures at points that would later become Roman towns. The case of Stanwick is germane because it was in the territory of the Brigantes, a tribe seemingly less well centralised than some of their southern counterparts. Furthermore, there are other instances of supposed 'sacred waterholes' in association with settlements in Yorkshire at

Heslington East (Ottaway, 2010) and Shiptonthorpe (Millett, 2006). Such local traditions may have played a role in the success of the Roman town at York, which was located at the flood-prone confluence between the Ouse and the Foss.

We know that to make it viable, water management in the Roman period involved similar sustained communal engagement; aqueducts, for instance, needed extensive maintenance, and this appeared to be the responsibility of local communities rather than central organising bodies. Even this book is somewhat guilty of concentrating on issues of defined power (whether local or incoming), and its projection through the urban landscape. It should be questioned, however, to what extent the creation of urban waterscapes was concerned with the formalisation of links between disparate communities in the wider landscape.

It is often reaffirmed how London's waterscape created a practical centrality that led to its primacy among the settlements of Roman Britain (Todd, 1989: 79; Perring, 1991: 1). The development of this waterscape may well have given the location a degree of symbolic centrality before the establishment of a defined town. The characteristics of the Thames at this point are seen as a key functionalist reason for the siting of a settlement. Yet how one rationalised that point of difference and change on the river was inevitably very different circa 2,000 years ago. London does represent the best bridging point and a place one can easily get to via wider river, sea and land routeways, but what this meant for people in the past is not easily explained. The Southwark islands with their traces of settlement and burial/funerary activity, plus the apparent ritual deposition of skulls in the Walbrook further back into prehistory, all hint at transient activity, bringing people together in this location for highly symbolic activities near to water. London was located close to the accepted borders of up to five Iron Age tribal territories. As evidenced in this book, the people of these surrounding territories appear to have placed an emphasis on sites located next to watery contexts. Moreover, movement of people between these territories may have already led to areas close to the eventual site of London being points of meeting and interchange, serving as a neutral place outside the control of any defined worldly authority. The Thames was seemingly a dividing line between the assumed territories of southern Britain, hence meeting within its influence may have been ideal for mediation of interterritorial issues and relations. With competing regional interests, it is likely that such a site would not be permanently occupied but still may have been intrinsically important and known (Bradley, 1990: 180–181).

The interpretations of the Tabard Square temple complex in Southwark as a religious focal point in a broader network of water-based sites perhaps contribute to this idea (Killock et al., 2015: 257), especially considering similar activity was happening nearby at Swan Street in a less monumental fashion, even in the first century. Many rural shrines, like Springhead, are assumed to have been important during prehistory but are given increased structural form in the Roman period as more people and ideas mixed. The sheer abundance of water at London is unrivalled in the other towns studied in this book. Accordingly, what must be emphasised is that the symbolic and ritual potential of London's waterscape already provided a platform to create a community, even without a defined local elite. The coming of Rome certainly expedites the situation, but the impact of these more ephemeral aspects of importance has almost certainly been under-played when evaluating the reasons why London rose to prominence.

What all this highlights is that the motivations for interacting with water were varied and complex at almost every site studied. At some sites we may see the motivations of incoming continental groups, at others there is a lingering sense of tradition, and at places like London there is a mixture of influences that are even harder to discern. Nonetheless, we can propose unique factors that make every waterscape a type of distinct hybrid of the urban form. Consequently, the construction of Roman urban water features can actually be justified from a whole range of perspectives, rather than merely monolithic aspiration to a unified Mediterranean ideal.

Stranger things: defamiliarising Roman urbanism in Britain

As can be seen in the case studies of Chapter 3, there is an intensification of interactions with water in many British towns during the first and second centuries, bringing the power of this medium further into the realm of sustained human experience. The relationship between 'human-made' and 'natural', practical and symbolic/religious was increasingly becoming blurred in relation to these waterscapes in the Late Iron Age. Human inter-vention was manipulating this medium in ways that better communicated meaning for local communities. In certain towns, like St Albans and Chich-ester, water may have been used to clarify the power of prominent indi-viduals and relate their authority to more ancient natural powers; in other places, like Canterbury, the interactions with water may have been part of regular activities that clarified relationships between a number of sur-rounding settlements. Increasingly, in these places the waterscapes were becoming entangled with the narrative of human experience. These changes were intensified and fortified by receptive incoming Roman ideas in the first and second centuries. The technological and architectural influ-ences of continental Europe gave increased ability to manipulate and interact with water, providing ways to strengthen previous associations with this medium in the landscapes of towns. Of course, along with these ideas came an influx of new people into the province with their own moti-vations for the manipulation of water in urban contexts. The argument made in this book is that the incoming 'Roman' beliefs surrounding water were not dissimilar from the indigenous Iron Age beliefs already present in Britain. Furthermore, exploration of Mediterranean examples shows a willingness to acknowledge a great deal of local diversification in relation to the specific conception of water in any given context.

The receptive nature of these incoming cultural influences creates the possibility, as shown above, to suggest mixing of beliefs and technologies surrounding water to express the particular social milieu of different settlements. The case studies explored in this book show hybrid waterscapes, without strict continuity from Iron Age contexts or a complete betrayal of previous associations with water as a result of Roman influences. This entanglement creates a situation where, despite divergent aims in different towns, the presence of water is increasingly centralised within settlements and is intimately related to the experience of place. The binding of meaning-laden water to urban space is a development that allows us to acknowledge further diversity and complexity in the construction of Roman towns. It is a way in which we can see the context of these towns having a direct impact on the way they were experienced; a context that is physical, social, political, religious and economic. In acknowledging this diversity, we can 'make strange' the Roman town, embracing a similar way of thinking to Richard Bradley (2007) and his discussions of the transformative connections of prehistoric monuments.

A number of examples of how this incorporation of water 'makes strange' an urban context can be drawn from this book. Wells, for instance, are shown to play a contributory role in such deconstruction of traditional and generalised interpretation. St Michael's Well at Lincoln is incorporated into both the fortress *principia* and the subsequent forum, with increasing elaboration as we move through the first and second centuries. The well's role as a central water focal point in a wider landscape where water is notably celebrated in both the Iron Age and the Roman period suggests that its association with the forum is far from coincidental. As shown above, the incorporation of the well may demonstrate important motives behind the construction of a town at this place. Taking these elements into consideration, this forum becomes involved in a local articulation of space, probably representing Roman manipulation of a symbolically rich waterscape in addition to a native tradition of increasing water interaction. Thus the function of the forum space is assuredly partly derived from this symbolic position in the landscape. The interpretation of the Upper City as being akin to a temple complex goes some way to confirming the diverging function of this particular forum from the generalised template of function often attributed to such features (Jones et al., 2003b: 7.2).

One may suggest similar transformative effects being derived from the wells in the towns of Dorchester and Silchester. Both have numerous examples of such features in their central monumental areas. In the case of

Dorchester, the pits appear to delineate ritual enclosures that likely had an impact on the way people experienced the monumental heart of the town. Silchester has a remarkable concentration of wells with votive offerings in the central insulae of the Roman town. As mentioned in the previous chapter, Fulford (2001) suggests that these features may have been a local tradition that continued and found new forms of expression in the settlement (as outlined previously, the deposition of different objects certainly changes over time). These wells/pits had the potential to influence the function and conception of surrounding urban structures. While archaeologists are fairly comfortable acknowledging wells as isolated instances of ritual veneration, the transport of their water to other structures or areas is given less thought. Moreover, their fundamental connection to a wider waterscape is underplayed. In this regard, wells in the Walbrook Valley may be more significant than previously acknowledged, possibly because they were seen as being connected to the meaning-laden waters of the nearby river. Even a well like the example we see in Building 8 at Verulamium, which never actually functioned correctly, could still have been fundamentally connected to wider beliefs and influence the perception of place (see Chapter 3).

This movement away from portrayal of water infrastructure as self-contained from the landscape has notable consequences. Aqueducts, along with their associated pipes and sewers, are often portrayed as the physical embodiment of sophisticated and practical incoming Roman power. Yet, as versed above, there is plenty of evidence suggesting that even in a Mediterranean setting they possessed wide-ranging symbolism predicated on the importance of the water they sourced in the wider landscape. In addition to their practical aspects, aqueducts transported the meaning inherent in spring and river sources into central urban locations. As versed in Chapter 2, the importance of the source was not eroded by the water's passage in a constructed conduit and thus could be a powerful influence in a new setting. If this was the case in the Mediterranean, it seems highly likely that in a province such as Britain, with its own diverse array of ritually important water sources, aqueducts cannot be satisfactorily explained in practical terms. This fundamentally challenges the conventional explanation of provincial aqueducts as the product of either 'Romans' constructing civilised works or 'natives' trying (and failing) to mimic a Mediterranean urban lifestyle (Kenyon, 1948; Stephens, 1985a, 1985b).

Instead, this book explores the idea that these structures were extensions of the meaning-laden waterscapes of many settlements in Britain. First and foremost, this anchors our interpretation into the local context. Yet it also gives us a convincing outlet for possible incoming motives, such as control and legitimisation, in addition to providing a platform for the increasing intensification of water control exhibited in the landscapes of many settlements throughout prehistory. Evaluating the aqueducts of

Britain in this fashion, we are not forced into portraying them as the poor relatives of their contemporary equivalents in continental Europe. Instead, we can outline ways that the more structurally simplistic aqueducts of Britain may well have been better suited to engage with the particular associations attributed to the waterscapes investigated. The relationship between the Dorchester aqueduct and the Poundbury hillfort, for instance, is perhaps referenced in the similar construction of the leat-style conduit and the surrounding earthworks. The proposed conduits at St Albans seemed to enhance connections with the River Ver, potentially building on the previously established dyke systems that appear to have symbolically linked locations to nearby rivers in addition to wider elements of reverence in the surrounding waterscape. At Lincoln future archaeological work may even confirm the radical proposal that the aqueduct siphoned water out of the town to the surrounding landscape, making it the strangest example of such a structure found associated with a major Roman town. Even in London, where there is less definitive evidence for an aqueduct, the Walbrook may have become akin to one, distributing water to many surrounding buildings in the monumental heart of the city, and in the process altering how people experienced them.

This discussion challenges the portrayal of municipal water in a direct sense, but also has a wider impact on the way we view the overall experience of towns. Water is such a malleable medium that its influence could permeate vast areas of the urban landscape. If aqueduct water supplied buildings, it is likely that this would have an effect on the way one experienced the building in question. Of all the urban beneficiaries of water, the bathhouses were clearly primary consumers of conduit-fed water in Britain. The channelling of water into these structures may have radically moved them away from their traditionally accepted function. If the Upper City bathhouse at Lincoln was supplied with water from the external Roaring Meg spring and its shrine, or from internal wells that possessed significance back into prehistory, this local meaning for the water could have defined how one would perceive the bathhouse.

It is, of course, undeniable that bathhouses possess similarities across the Roman world. However, we have become far too invested in the process of making these similarities their defining aspect. The ideas of Richard Hingley (2005) on the topic of globalisation could be more thoroughly applied here; similar form does not necessitate identical function in a globalised world. One might even venture the radical idea that modern archaeologists are projecting similarity that is not present in the evidence. After all, the examples we study in this book rarely share identical spatial dimensions, distribution of rooms, positioning in towns or water supply. Moreover, in the classical sources there are arrays of differences that characterise each bathhouse – and that is without even considering the context of associations with water at the sites discussed in Chapter 3. Once we do this, we are left with a structure that can only be fully understood via analysis at the local level. For example, the bathhouses at Dorchester, Lincoln, Silchester, St Albans and Wroxeter have all been approached at one point in similar traditional terms. We have labelled their rooms and outlined how individuals would go about performing the classical bathing routine. Yet the evidence of Chapter 3 displays how each of these bathhouses have contexts that appear to be far more influential in the experience of these spaces. Accepting this makes it far harder to create a neat narrative of place, but it contributes to a far more considered sense of agency in the creation of these 'Roman' buildings.

A major result of the analysis of various parts of the urban waterscape explored throughout this book is that it becomes necessary to acknowledge a great deal of complexity in the interpretation of multiple features of the Roman towns investigated. It illustrates a series of urban settings that were formed and developed with numerous influences and motivations, both local and external. The common characteristics and functions attributed to wells, aqueducts, sewers and bathhouses are shown to be limited, even from the perspective of incoming Mediterranean influences. These features have been 'made strange' by a consideration of how they related to their immediate context, rather than the extent to which they conformed to an illusory archetypal 'Roman' form and function. This, in turn, has a major effect on the way we view Roman towns as a whole. Water is such an integral part of the traditional idea of what constituted a Roman settlement. If its role was markedly more complex, then the development of the town probably mirrored this hybrid identity. A more questioning approach to the broad familiarity of Roman towns in Britain can thus help us gain a deeper understanding of how they were experienced. Moreover, such a strategy would provide more thorough local 'ownership' of archaeological heritage, outlining ways that settlements adapted and changed in relation to their unique contexts.

Changing environmental conditions and urban waterscapes

Outside the narrow confines of Roman archaeology, work in the last 20 years to study the climate of the Roman period has given us another layer of understanding to human engagement with the natural world. Through combined analysis of proxy sources (including tree rings, ocean sediments, lake sediments, historical documents, etc.) there is a current consensus that the period from the first through to the third century represented a time of remarkable climate stability for Western Europe, with temperatures potentially reaching historic highs (Ljungqvist, 2010: 347). Indeed, the 'Roman Warm Period', as it has become known, created the ideal circumstances for agricultural expansion, with consistent weather patterns enabling a reliable relationship between people and their local landscapes. There has been much speculation on how the interruption of this stability, partly through changes in the North Atlantic oscillation, could have led to catastrophic events in the history of the Roman world; environmental pressures, for instance, seemed to coincide with the crisis of the third century and the tumultuous fifth-century migration period (Drake, 2017; Büntgen *et al.*, 2011).

These studies provide us with another layer of diversity in the reasoning behind the uptake of more complex water infrastructure. Instead of a uniform cultural directive, one can posit that the favourable conditions of the first and second centuries meant this was an ideal time for water infrastructure to expand on an unprecedented scale, regardless of imperial influences. There are obvious dangers in trying to make grand assumptions about community reactions to climate; for one, they echo some of the unhelpful reductionism of processualists discussed earlier in this book (Izdebski *et al.*, 2016: 11). Moreover, the accuracy of such reconstructions is continually being re-evaluated (Ljungqvist, 2010: 339). Nevertheless, it is fair to say that such stable conditions were likely conducive to the communal water projects highlighted in previous chapters. In Britain, by the Late Iron Age there had already been marked intensification in the engagement with waterscapes at many locales. We should appreciate how the circumstances of the regional climate could have strengthened local belief systems involving water.

From the perspective of people living at this time, there was a real negotiation with nature to secure such prosperous conditions; one had to placate and satisfy a deified natural world. Accordingly, this increasing entanglement with water was actively producing even more favourable conditions from the point of view of people in Roman Britain. Thus even if they do not seem to be practical actions to us, such initiatives could have been reinforced as strong outlets for community time. Moreover, this circle of dependency would have laid the platform for specific individuals and groups to cast themselves as facilitators and providers of prosperity (see O'Sullivan, 2008: 48); perhaps akin to the activity highlighted at St Albans or Chichester.

Consideration of these factors has a role to play in how we interpret reactions to the urbanisation process in provinces like Britain. At Lincoln it was highlighted how previous accounts emphasised the Roman developments as an overt appropriation of an important local ritual site (see Jones, 2003; Rogers, 2013). In this book it is argued that the reality may have been a great deal more nuanced. An aspect of that complexity may well have been the stable climate conditions. There was unprecedented human development of the waterscape in the Roman period, and this could have initially been interpreted as intruding on sacred territory. If this had been followed by a dramatic change in agricultural conditions, for instance, it could have reinforced local resistance to changes. However, environmentally stable and productive conditions are perhaps more likely to have led to tolerance and legitimisation of these actions.

These paleoclimate studies do hint at an eventual change in this stability, which has been said to coincide with the turbulent events of the later Empire (Drake, 2017: 2). It is possible that some of the reaction to these changing times can be witnessed in the ways that people interact with water, particularly on a religious level, in Britain. Mithraism and the eventual rise of Christianity are both mentioned a number of times in the survey of towns in Chapter 3. They are religions with clear connections to water – Mithras was even born from a rock, like a spring – and both become dominant parts of the mythology of waterscapes in the later Roman Empire. There is a myriad of reasons for their popularity, but one can speculate that the changes in climate stability in a place like Britain could have played a part in their appeal to different communities.

While the prosperity of the first and second centuries likely reinforced beliefs around water, changing conditions could well have started to undermine long-lived practices and motivated different types of engagement. As noted in Chapter 3, there are more lavish displays of votive offerings in wells from places like London and Silchester from the third century onwards, and the Temple of Mithras was established in the centre of the ritually important Middle Walbrook area in London. These practices might be the result of more trying times and a different, more personal, appeal to the gods. In fact, a far more private and restricted relationship to water was manifest in the practices of both Mithraism and Christianity, whether in the secret rites of Mithraism or the close control of water infrastructure by the Church in Late Antiquity and beyond. Certainly, the waterscapes of Canterbury and Cirencester are both defined by their interaction with Christian abbeys in the medieval period (Reece and Broxton, 2011).

If climate unpredictability created a greater strain on individuals and family groups to provide for themselves, the complex gestures of public water engagement witnessed in many places in Britain from prehistory through the High Roman period may have become unsustainable. In fact, such changes in habits may have had a circuitous negative effect on the people living in towns. Dependencies were inevitably created between people and their local waterscapes in the first and second centuries, and small alterations in the upkeep of systems could have made the effects of unfavourable climate conditions even more acute. There has been research into such processes in other provincial territories. At the city of Leptis Magna, briefly mentioned in Chapter 2, reclamation and damming of the local wadi created the opportunity for greater engagement with water in the town. However, the lack of maintenance of the dam from the third century onwards eventually led to even more damaging floods (Pucci *et al.*,

2011: 182–183). British systems may not have been as monumental, but we still see potential effects – such as the widespread flooding in Canterbury during the third century (Rogers, 2013).

Of course, such interpretations are speculative and would need more thorough examination of these later periods, which are not the focus of this book. However, the relationship of people to their urban waterscapes had undoubtedly changed by the post-Roman period, and climate may have been a principal factor in that development. This is another marker of how analysis of the Roman period, and the overall message of difference and variability in the past, can help us break the compulsion to enforce our status quo. Climate change required adaption of the relationship to water in this period, and the more profound challenges we are likely to face in the twenty-first century may well require something even more radical. Of course, to achieve accurate analysis of such issues we need greater future cooperation between archaeologists and specialists working in other related disciplines, like environmental scientists (see Izdebski *et al.*, 2016).

Water and the identity of our urban future

One of the vital aspects of embracing a 'strange' and hybrid Roman-period approach to water is that we underline this sense of change and difference to the past. What can be seen in the studies in this book is that each site had unique reasons and solutions for bringing water into central urban areas. While some of these examples share similarities to practices in the Iron Age and others seem to bear more comparison to contemporaraneous developments in continental Europe, they all represent a human engagement with water that differs from what was present before or after. This is not the beginning of an inexorable teleological march to the 'civilised' modern water infrastructure of today, as past accounts have frequently implied. In fact, the way water was understood and valued in the towns of Roman Britain is vastly different from its treatment in the twenty-first-century settlements that now overlie them. There are similarities, like those mentioned above concerning Iron Age and Mediterranean features, but it is the differences seen at the local level that appear to have had the most notable impact on the way people interacted with water.

As noted in Chapter 1, that would be a reasonably unremarkable comment if we were dealing with prehistoric evidence. However, Roman archaeology so often finds its relevance with a wider audience through avenues of familiarity. To some extent that is inevitable, and it can be a positive way to get individuals to care about heritage issues. But, as is the case with technological systems like Roman water supply, we can also start using that perceived cultural proximity to legitimise the maintenance of an unhealthy status quo. The perception of Roman aqueducts or bathhouses sharing a similar utility to modern water pipes or spas lends the weight of historical precedent to these current engagements with water. This is a particularly pertinent issue when change is required, as is the case with our current relationship to water and the natural world in general.

Water scarcity and surplus are emerging as prime symptoms of climate change, and will prove a challenge to urban centres throughout the world over the next century (see Sedlak, 2015). Changes need to be made in the way we relate to and experience water within our cities and their hinterlands. Yet there is widespread indifference to this issue among sections of the general public. Even in areas of drought there can be resistance to any changes to long-held habits – the recent droughts in California and Cape Town (South Africa) are cases in point. The way we perceive historical precedent plays a role in this naivety. By overplaying the connections between twentieth-century urban aims and those of the ancient world, we promote the idea of inherited behaviour and the continuance of a tradition. These are universally powerful concepts that justify maintenance of an untenable status quo and make it hard for people to accept change away from it, whether that is because of fear or a sense of entitlement, or perhaps both.

As a result, the ideas of 'difference' and contrast between the twenty-first century and the Roman past expressed in this book could have more relevance for broader progressive approaches to the future of urbanism. First, in a straightforward sense, acknowledging change in the past rather than continuity communicates the likelihood that future developments will be different from our current situation. Instead of using evidence from antiquity to bolster outdated attitudes, the 'strange' Roman municipal water presented in this book could help us engage with the stranger-still future of urbanism. The consideration of hybridity, for instance, speaks to current thinking on this topic. As explained above, part of the hybridity we can observe in Roman Britain concerns the balance between human and natural aspects of water supply. The previous chapters are a testament to the fact that in these settlements of the past there was far less separation between these two terms. Water was a powerful and space-defining element of life, and its incorporation into various parts of the urban fabric was highly significant. Out of necessity and strength of belief, the natural world was not the marginalised and subservient aspect of existence that it is today.

This everyday awareness of water and the wider natural world is probably not achievable today, but it is easy to see how it engages with current popular movements in conservation and urban planning. For example, the 'rewilding' movement has gained increasing traction over the last ten years. It describes a process by which we can we can redress the balance between human development and the natural ecosystems within which it occurs. These proposals often take the form of a reintroduction of various plant and animal species to create more robust and diverse wildlife that can resist the damaging effects of human-driven climate change (see Fraser, 2009; Monbiot, 2014). The rewildingbritain.org.uk website summarises this process as 'bringing nature back to life' and 'restoring living systems'. It is easy to see how this type of thinking meshes with what is observed in the case studies of this book. In Roman Britain we have a clearly 'living' nature that was highly variable and interconnected with human issues of politics, religion, economics, health and more besides. The intensification of engagement with water that we see in the Roman period integrates this natural force into the fabric of daily life to an extent hitherto not witnessed in the province. Furthermore, the act of reintroducing species and altering the destiny of their habitats is a tacit acknowledgement of a hybrid relationship between what is considered human and what is natural.

The other vital aspect of the hybridity presented in this account is the local diversity in the archaeology of Roman Britain. This is a theme in Robert Macfarlane's book *Landmarks*, discussing the rapidly declining vocabulary that English speakers use to describe the natural world. He notes that in the recent past in Britain there were a plethora of local terms to describe landscape features. For instance, terms describing wells in Cornwall alone included 'peath', 'fenton' and 'shute' (Macfarlane, 2015: 133). This study raises serious questions about the past and the assumptions we make, even by applying our modern language. Macfarlane's book is a reaction to a world where the diversity of our language relating to nature is becoming increasingly limited. He styles it a 'counter desecration phrasebook' (ibid.: 15); something that can be used as a manual to combat the disappearance of 'nature words' from even the *Oxford Junior Dictionary*. At its heart, however, Macfarlane is essentially 'making strange' the past, creating a window on a time when language hard-wired local variability into the perception of landscape features. Considering Macfarlane's timescale is more recent, when the concept of Britain was more culturally and linguistically unified, it might be expected that the Roman period would present greater diversity. This, of course, is one of the primary efforts of this book: to move beyond our simplistic labelling of water structures and integrate them into their landscape context. This may give it a degree of relevance in understanding the complex relationship we will have with nature moving forward.

However, the issues raised in this book share common ground with studies of water outside the cultural sphere of the humanities. The 2015 book *Water 4.0*, written by the environmental engineer David Sedlak (University of Berkley), gives a detailed analysis of the historical development of water management infrastructure, particularly during modernity. Sedlak's primary goal was to provide an evidence-based assessment of the likely form that our water infrastructure systems will have to take in the face of the pressures of climate change. The conclusion he reaches for the nature of this so-called 'Water 4.0' is significantly different from the solutions of the last century. While we have become comfortably naive about our centralised water systems, in many places they are not prepared for the dramatic impact of environmental changes. In Sedlak's view, one of the radical changes that may be required is the transfer of 'responsibility for acquisition, treatment, and management back to the household or neighbourhood' (ibid.: 278). Ultra-efficient household appliances and elimination of wasteful outdoor water use may be a start to such a movement. What is clear is that even if we try to pursue the modernisation of old centralised systems, a large fraction of the population will have to be involved, and therefore we will need to overcome 'the public's continuing desire to ignore water management' (ibid.: 279).

As is rehearsed copiously above, Roman water management is often a historical bulwark to the primacy of these (now outdated) centralised urban water systems in our cities. However, the case studies analysed in this book should give us an inkling of how they could be relevant in formulating a view of possible divergent futures, similar to those outlined by Sedlak. We can see at all these sites how local decisions were at the heart of the specific form of urban water system. There was no 'one-size-fits-all' solution that could be imprinted on a new site. There had to be an awareness of the particular attributes of the local landscape and waterscape, and how people related to them. Moreover, when these urban water networks were established in Britain, local populations likely played a significant role in maintaining them. People were invested in these systems, and not just because they provided drinking water. This connection encompassed expressions of group identity, religious beliefs, political statements and perceptions of time or seasonality. Rather than endorsing a universal practical approach, even the centralising force of the Roman army appears to be as concerned with the local impact of water systems.

If we are to combat the challenges of the coming years, people need to be similarly knowledgeable about their local environment, willing to dedicate their own time to the upkeep of new systems, passionate about their role in a communal project and aware of their place in a broader changing political landscape. So while it is common to suggest that the organisation of urban water in the Roman period is a formative development stage of the municipal water management of the last 150 years, it might be more accurate to see it as an inspiration for the radically different challenges of the next 150 years.

Conclusions

The outdated parallels between the Roman period and our modern world still endure in the interpretation of towns. Today archaeologists may be more concerned with highlighting the role of indigenous people in provinces like Britain than in the past, but the contribution of these people to the development of urban space has not extended to the manipulation of the 'well-understood' building types that have come to define Roman settlements. We also continually underplay the extent to which incoming Roman influences could be receptive to the integration of particular customs and beliefs found in provincial contexts. This propagates the idea of a clear-cut divide between local people and incoming Romans: we can acknowledge straight continuity with Iron Age practice appearing in these towns, but characteristically Roman structures are still understood as being built for universal Roman reasons and experienced in a familiar Roman fashion. Such easily defined and generalised motivations are likely to be unrepresentative of the complexity of urban development during this period in Britain. The alternative, suggested in this book, is perceiving Roman towns as places characterised by hybrid influences, with a combination of incoming and local beliefs providing the impetus for development and experience.

Water was selected as the medium for this study because it shows the drawbacks of past research while illustrating the potential of considering hybridity. Within the popular consciousness there continues to be a stronger connection between waterscapes of Roman towns and the modern world than with those of pre-Roman Britain. In reality this disparity is probably based more on the bias of our approach than the actual evidence. Exploration in this book shows that even in the core Roman cultural zone of the Mediterranean, urban water supply possessed symbolic and ritual associations; but these special values were very much dependent on the local context and the particular attributes of its waterscape. This suggests that in Roman Britain the incoming influences were probably subject to manipulation and variability based on the incorporation of local, provincial associations with water, thereby creating hybridity within towns.

This final chapter demonstrates the extent to which we can detect differences in the development of hybrid waterscapes in the various towns of Britain. It gives us a picture of the variability while also underlining the multidimensional meaning that waterscapes could have possessed in any one location. By applying this approach we get a more definite sense of the variable ways in which people could have justified different Roman-period features, based on their own experiences. This portrays a pronounced local agency ingrained in the identity of urbanism in Roman Britain. Water can seem like the most familiar of substances: something with which we must interact daily, and which has a value for all life. This fundamental importance, paired with its physical juxtaposition of malleability and destructive force, creates a platform for complex meaning and reception throughout history. We should be open to the variable interpretations of archaeological evidence that this can create, forcing us to acknowledge a stranger picture of Roman urbanism, less connected to the present but more rooted in its local context. Thus while the study of water supply can appear to be a well-understood trope of Roman archaeology, there is much we are yet to appreciate about the effect these systems had on the experience of Roman urbanism.

Note

1 It is worth mentioning that Legio IX, which is deemed responsible for much building at Lincoln, had been stationed in areas within temperate Europe for many years prior to its appearance in Britain (Jones *et al.*, 2003b: 6.9).

References

Abell, E. and J. Chambers (1971) *The Story of Lincoln*, 2nd edn. Wakefield: S. R. Publishers.

Ackroyd, P. (2008) *Thames: Sacred River*. London: Vintage.

AE (1889–) *L'annee epigraphique*. Paris: Presses Universitaires de France.

Aldhouse-Green, M. J. (1986) *The Gods of the Celts*. Gloucester: A. Sutton.

Aldrete, G. S. (2007) *Floods of the Tiber in Ancient Rome*. Baltimore, MD: Johns Hopkins University Press.

Allinne, C. (2007) 'Les villes romaines face aux inondations. La place des données archéologiques dans l'étude des risques fluviaux', *Géomorphologie : Relief, Processus, Environnement*, 13(1), available at http://journals.openedition.org/geomorphologie/674.

Allison, P. M. (2007) 'Domestic spaces and activities', in J. J. Dobbins and P. W. Foss (eds) *The World of Pompeii*. Oxford: Routledge, 269–279.

Ammerman, A. (1990) 'On the origins of the Forum Romanum', *American Journal of Archaeology*, 94, 627–645.

Andrews, P., E. Biddulph, A. Hardy, R. Brown and R. Goller (2011) *Settling the Ebbsfleet Valley High Speed 1 Excavations at Springhead and Northfleet, Kent: The Late Iron Age, Roman, Saxon and Medieval Landscape*, Vol. 1. Salisbury: Oxford Wessex Archaeology.

Anthony, I. (1970) ' "St. Michaels", St. Albans, excavations 1966', *Hertfordshire Archaeology*, 2, 51–61.

Arnold, B. (1999) ' "Drinking the feast": Alcohol and the legitimation of power in Celtic Europe', *Cambridge Archaeological Journal*, 9, 71–93.

Ashby, T. (1935) *The Aqueducts of Ancient Rome*. Oxford: Clarendon Press.

Ashby, T., A. E. Hudd and F. King (1904) 'VII – Excavations at Caerwent, Monmouthshire, on the site of the Romano-British city of Venta Silurum, in the years 1901 and 1903', *Archaeologia or Miscellaneous Tracts Relating to Antiquity*, 59(2), 87–124.

Ashby, T., A. E. Hudd and F. King (1909) 'XXI – Excavations at Caerwent, Monmouthshire, on the site of the Romano-British city of Venta Silurum, in the years 1907 and 1909', *Archaeologia or Miscellaneous Tracts Relating to Antiquity*, 61(2), 565–582.

Ashby, T., A. E. Hudd and F. King (1911) 'XIX – Excavations at Caerwent, Monmouthshire, on the site of the Romano-British city of Venta Silurum, in the years

1909 and 1910', *Archaeologia or Miscellaneous Tracts Relating to Antiquity*, 62(2), 405–448.

Atkinson, M. and S. J. Preston (1998) 'The Late Iron Age and Roman settlement at Elms Farm, Heybridge, Essex, excavations 1993–5: An interim report', *Britannia*, 29, 85–110.

Baker, P. A. (2011) 'Collyrium stamps: An indicator of regional practices in Roman Gaul', *European Journal of Archaeology*, 14(1/2), 158–189.

Ball, P. (2016) *The Water Kingdom: A Secret History of China*. London: Bodley Head.

Barfield, L. and M. Hodder (1987) 'Burnt mounds as saunas, and the prehistory of bathing', *Antiquity*, 61(233), 370–379.

Barker, P. and M. Armour-Chelu (1997) *The Baths Basilica Wroxeter: Excavations 1966–90*. London: English Heritage.

Barker, P. and G. Webster (1990) *From Roman Viroconium to Medieval Wroxeter: Recent Work on the Site of the Roman City of Wroxeter*. Worcester: West Mercian Archaeological Consultants.

Barolsky, P. (2005) 'Ovid, Bernini, and the art of petrification', *Arion*, 13(2), 149–162.

Bates, A. (2018) Personal communication.

Bayley, J., B. Croxford, M. Henig and B. Watson (2009) 'A gilt-bronze arm from London', *Britannia*, 40, 151–162.

Beasley, M. (2006) 'Roman boundaries, roads and ritual: Excavations at the Old Sorting Office, Swan Street, Southwark', *Transactions of the London and Middlesex Archaeological Society*, 57, 23–68.

Bedwin, O. and C. Orton (1984) 'The excavation of the eastern terminal of the Devil's Ditch (Chichester Dykes), Boxgrove, West Sussex, 1982', *Sussex Archaeological Collections*, 122, 63–74.

Benfield, S. F. and S. Garrod (1992) 'Two recently-discovered Roman buildings in Colchester', *Essex Archaeology & History*, 23, 25–38.

Bennett, P. (1980) 'Excavations at No. 3 Beer Cart Lane', *Archaeologia Cantiana*, XCV, 270–272.

Bennett, P. (1987) 'The conduit house: The cathedral water supply', in *Canterbury Excavations Intra- and Extra-Mural Sites, 1949–55 and 1980–84*. Maidstone: Kent Archaeological Society, 25–33.

Bennett, P. (1990) 'St Augustine's water supply', in *Canterbury's Archaeology 1988–1989*. Canterbury: Canterbury Archaeological Trust, 13.

Bidwell, P. T. and D. M. Bailey (1979) *The Legionary Bath-house and Basilica and Forum at Exeter*. Exeter: Exeter City Council and University of Exeter.

Bird, J. (1996) 'Frogs from the Walbrook: A cult pot and its attribution', in J. Bird, M. Hassall and H. Sheldon (eds) *Interpreting Roman London: Papers in Memory of Hugh Chapman*. Oxford: Oxbow, 119–127.

Blackbourn, D. (2007) *The Conquest of Nature: Water, Landscape and the Making of Modern Germany*. New York: Norton.

Blagg, T. F. C. (1990) 'Architectural munificence in Britain: The evidence of inscriptions', *Britannia*, 21, 13–31.

Blair, I. (2002) 'Roman London's waterworks: The Gresham Street discoveries', *Current Archaeology*, 180, 509–516.

Blair, I. and J. Hall (2001) 'Water management in Roman London', *Minerva*, 12(6), 4.

Blair, J. (2007) *The Church in Anglo-Saxon Society*. Oxford: Oxford University Press.

Blockley, K., A. P. Detsicas and J. Elder (1995) *The Archaeology of Canterbury*, Vol. 5. Canterbury: Canterbury Archaeological Trust.

Bogaers, J. E. A. T. (1979) 'King Cogidubnus in Chichester: Another reading of RIB 91', *Britannia*, 10, 243–254.

Boon, G. C. (1974) *Silchester, the Roman Town of Calleva; with a Folding Plan of the Town Showing Details Traced from Aerial Photographs, 40 Photographs and 42 Drawings*. Newton Abbot: David & Charles.

Boutwood, Y. (1998) 'Prehistoric linear boundaries in Lincolnshire and its fringe', in R. Bewley (ed.) *Lincolnshire's Archaeology from the Air*, Occasional Papers in Lincolnshire History and Archaeology 11, 29–46.

Bradley, R. (1971) 'A field survey of the Chichester Entrenchments', in B. Cunliffe (ed.) *Excavations at Fishbourne 1961–1969*. London: Society of Antiquaries, 17–37.

Bradley, R. (1990) *The Passage of Arms: An Archaeological Analysis of Prehistoric Hoards and Votive Deposits*. Cambridge: Cambridge University Press.

Bradley, R. (1993) *Altering the Earth: The Origins of Monuments in Britain and Continental Europe. The Rhind Lectures 1991–92*. Edinburgh: Society of Antiquaries of Scotland.

Bradley, R. (2000) *An Archaeology of Natural Places*. London: Routledge.

Bradley, R. (2007) 'Making strange: Monuments and the creation of the earlier prehistoric landscape', in A. J. Fleming and R. Hingley (eds) *Prehistoric and Roman Landscapes: Landscape History after Hoskins*. Macclesfield: Windgather Press, 33–43.

Bradley, R. and Gordon, K. (1988) 'Human skulls from the River Thames, their dating and significance', *Antiquity*, 62(236), 503–509.

Breeze, A. (2002) 'Does Corieltavi mean "army of many rivers"?', *Antiquaries Journal*, 82, 307–309.

Brett, M. and M. Watts (2008) 'Excavations at Stepstairs Lane, 2002–3', in N. Holbrook (ed.) *Excavations and Observations in Roman Cirencester, 1998–2007: With a Review of Archaeology in Cirencester 1958–2008*. Cirencester: Cotswold Archeology, 70–83.

Brewer, R. (2006) *Caerwent Roman Town*. Cardiff: Cadw.

Brigham, T (2001) 'The Thames and the Southwark waterfront in the pre-Roman period', in B. Watson, T. Brigham and T. Dyson *London Bridge: 2000 Years of a River Crossing*. London: Museum of London Archaeology Service, 8–12.

Britton, J. (1812) *The Beauties of England: Lincolnshire*. London: Thomas Maiden.

Brooks, H. and P. Crummy (1984) 'Excavations at Middleborough 1979', in P. Crummy (ed.) *Excavations at Lion Walk, Balkerne Lane, and Middleborough, Colchester, Essex*, Colchester Archaeological Report 3. Colchester: Colchester Archaeological Trust.

Brooks, H., W. Clarke, M. Górniak and L. Pooley (2009) *CAT Report 347: Roman Buildings, the Rear Face of the Roman Town Wall and Archaeological Investigations in Insulas 1a, 1b, 9a and 9b, at the Sixth Form College, North Hill, Colchester, Essex: April 2005–March 2006*. Colchester: Colchester Archaeological Trust.

Bruchac, J. (1993) *The Native American Sweat Lodge: History and Legends*. Freedom, CA: Crossing Press.

Bruneton, H., G. Arnaud-Fassetta, M. Provansal and D. Sistach (2001) 'Geomorphological evidence for fluvial change during the Roman period in the lower Rhone Valley (southern France)', *Catena*, 45(4), 287.

Bruun, C. (2012) 'Roman emperors and legislation on public water use in the Roman Empire: Clarifications and problems', *Water History*, 4(1), 11–35.

Bryant, S. (2007) 'Central places or special places? The origins and development of "oppida" in Hertfordshire', in C. Haselgrove and T. Moore (eds) *The Later Iron Age in Britain and Beyond*. Oxford: Oxbow Books, 62–81.

Bryce, J. (1914) *The Ancient Roman Empire and the British Empire in India: The Diffusion of Roman and English Law Throughout the World; Two Historical Studies*. Oxford: Oxford University Press.

Budd, P. and T. Taylor (1995) 'The faerie smith meets the bronze industry: Magic versus science in the interpretation of prehistoric metal-making', *World Archaeology*, 27(1), 133.

Büntgen, U., W. Tegel, K. Nicolussi, M. McCormick, D. Frank, V. Trouet, J. O. Kaplan, F. Herzig, K.-U. Heussner, H. Wanner, J. Luterbacher and J. Esper (2011) '2500 years of European climate variability and human susceptibility', *Science*, 331(6017), 578–582.

Burgers, A. (2001) *The Water Supplies and Related Structures of Roman Britain*. Oxford: J. & E. Hedges.

Bushe-Fox, J. P. (1916) *3rd Report on the Excavations on the Site of the Roman Town at Wroxeter, Shropshire, in 1914*, Vol. 4. London: Society of Antiquaries.

Caesar, Julius (trans. H. J. Edwards, 1917) *Bellum Gallicum (The Gallic Wars)*. London: Heinemann.

Celsus (trans. W. G. Spencer, 1938) *De medicina (On Medicine)*. Cambridge, MA: Harvard University Press.

Chadwick, A. M. (2007) 'Trackways, hooves and memory-days – Human and animal movements and memories around the Iron Age and Romano-British rural landscapes of the English north Midlands', in V. Cummings and R. Johnston (eds) *Prehistoric Journeys*. Oxford: Oxbow Books, 131–152.

Chanson, H. (2004) 'Hydraulics of rectangular dropshafts', *Journal of Irrigation and Drainage Engineering*, 130(6), 523–529.

Chapman, E. M., F. Hunter, P. Wilson and P. Booth (2012) 'I. Sites explored', *Britannia*, 43, 271–354.

Charles, B. M., A. Parkinson, S. Foreman and C. Allen (2000) 'A Bronze Age ditch and Iron Age settlement at Elms Farm, Humberstone, Leicester', *Transactions of the Leicestershire Archaeological and Historical Society*, 64, 113–220.

Charlesworth, D. (1972) *The Glass*. London: Society of Antiquaries, 196–215.

Cicero (trans. C. D. Yonge, 1878) *De Legibus (On the Laws)*. London: G. Bell.

CIL (1862–) *Corpus Inscriptionum Latinarum*. Berlin: Reimer.

Clarke, J. R. (1991) *The Houses of Roman Italy, 100 B.C–A.D. 250: Ritual, Space, and Decoration*. Berkeley, CA: University of California Press.

Clay, P. and R. Pollard (1994) *Iron Age and Roman Occupation in the West Bridge Area, Leicester: Excavations 1962–1971*. Leicester: Leicester Arts and Museums Service.

Coarelli, F. (1986) *Il Foro Romano: Periodo Arcaico*. Rome: Quasar.

Coles, J. (2001) 'North European bronzes, rock art and wetlands: Looking for context and relations. A preliminary study', in B. Purdy (ed.) *Enduring Records: The Environmental and Cultural Heritage of Wetlands*. Oxford: Oxbow Books, 148–157.

Collie, N. (2013) 'Ritualising encounters with subterranean places: An investigation of urban depositional practices of Roman Britain', PhD thesis, University of Tasmania.

Cooper, N. J. and R. Buckley (2003) 'New light on Roman Leicester (Ratae Corieltauvorum)', in J. Wacher and P. R. Wilson (eds) *The Archaeology of Roman Towns: Studies in Honour of John S. Wacher*. Oxford: Oxbow Books, 31–43.

Cooper, N. J., G. Campion and P. Lowther (2006) *The Archaeology of the East Midlands: An Archaeological Resource Assessment and Research Agenda*. Leicester: University of Leicester Archaeological Services.

Cordner, W. S. (1946) 'The cult of the Holy Well', *Ulster Journal of Archaeology*, 9, 24–36.

Corpus Agrimensorum Romanorum (trans. B. Campbell, 2000). London: Society for Promotion of Roman Studies.

Cowan, C. (2009) *Roman Southwark Settlement and Economy: Excavations in Southwark, 1973–1991*. London: Museum of London Archaeology.

Creighton, J. (2000) *Coins and Power in Late Iron Age Britain*. Cambridge: Cambridge University Press.

Creighton, J. (2001) 'The Iron Age–Roman transition', in S. James and M. Millett (eds) *Britons and Romans: Advancing an Archaeological Agenda*. York: Council for British Archaeology, 4–12.

Creighton, J. (2006) *Britannia: The Creation of a Roman Province*. London: Routledge.

Cromer, E. (1910) *Ancient and Modern Imperialism*. London: John Murray.

Crummy, N. (2006) 'Worshipping Mercury on the Balkerne Hill, Colchester', in P. Crummy and P. Ottaway (eds) *A Victory Celebration: Papers on the Archaeology of Colchester and Late Iron Age–Roman Britain Presented to Philip Crummy*. Colchester: Friends of Colchester Archaeological Trust, 55–69.

Crummy, N. and C. Pohl (2008) 'Small toilet instruments from London: A review of the evidence', in J. Clark, J. Cotton, J. Hall, R. Sherris and H. Swain (eds) *Londinium and Beyond: Essays on Roman London and its Hinterland for Harvey Sheldon*. York: Council for British Archaeology, 212–224.

Crummy, P. (1984) *Excavations at Lion Walk, Balkerne Lane, and Middleborough, Colchester, Essex*. Colchester: Colchester Archaeological Trust.

Crummy, P. (1988) 'Colchester (Camulodunum/Colonia Victriciensis [sic])', in G. Webster (ed.) *Fortress into City: The Consolidation of Roman Britain, First Century AD*. London: Batsford, 22–48.

Crummy, P. (1991) 'Where was Colchester's public bath-house?', *Colchester Archaeologist*, 4, 8–12.

Crummy, P. (1997) *City of Victory: The Story of Colchester – Britain's First Roman Town*. Colchester: Colchester Archaeological Trust.

Crummy, P. (2005) 'The circus at Colchester (Colonia Victricensis)', *Journal of Roman Archaeology*, 18, 267–277.

Cunliffe, B. W. (1971) *Excavations at Fishbourne, 1961–1969*. London: Society of Antiquaries.

Curwen, E. C. (1954) *The Archaeology of Sussex*. London: Methuen.

Dark, K. R. (1993) 'Town or temenos? A reinterpretation of the walled area of Aqua Sulis', *Britannia*, 24, 254–255.

Darling, M. J. and M. J. Jones (1988) 'Early settlement at Lincoln', *Britannia*, 19, 1–57.

Davies, J. A. (2001) *Venta Icenorum: Caistor St. Edmund Roman Town*. Dereham: Norfolk Archaeological Trust.

Davies, J. A. (2009) *The Land of Boudica: Prehistoric and Roman Norfolk*. Oxford: Oxbow Books.

Davies, P. and J. Robb (2002) 'The appropriation of the material of places in the landscape: The case of tufa and springs', *Landscape Research*, 27(2) 181–185.

de Kleijn, G. (2001) *The Water Supply of Ancient Rome: City Area, Water, and Population*. Amsterdam: Gieben.

DeLaine, J. (1996) 'De aquis suis? The commentarius of Frontinus', in C. Nicolet (ed.) *Les littératures techniques dans l'Antiquité romaine: statut, public et destination, tradition: sept exposés suivis de discussions, Vandoeuvres-Genève, 21–25 août 1995*. Geneva: Fondation Hardt, 117–145.

DeLaine, J. (1999a) 'Benefactions and urban renewal: Bath buildings in Italy', in J. DeLaine and D. Johnston (eds) *Roman Baths and Bathing: Proceedings of the First International Conference on Roman Baths Held at Bath, England, 30 March–4 April 1992*. Portsmouth, RI: Journal of Roman Archaeology Supplementary Series 37, 67–75.

DeLaine, J. (1999b) 'Introduction: Bathing and society', in J. DeLaine and D. Johnston (eds) *Roman Baths and Bathing: Proceedings of the First International Conference on Roman Baths Held at Bath, England, 30 March–4 April 1992*. Portsmouth, RI: Journal of Roman Archaeology Supplementary Series 37, 7–17.

Derks, T. (1998) *Gods, Temples, and Ritual Practices: The Transformation of Religious Ideas and Values in Roman Gaul*. Amsterdam: Amsterdam University Press.

de Ruyt, C. (1983) *Macellum: Marché alimentaire des romains*. Louvain-La-Neuve: Institut supérieur d'archéologie et d'histoire de l'art, College Érasme.

di Capua, F. (1940) 'Appunti su l'origine e sviluppo delle terme romane', *Accademia di Architettura, Lettere e Belle Arti*, 20, 81–160.

Dionysius Halicarnassensis (trans. E. Cary, 1945) *Antiquitatum Romanarum (Roman Antiquities)*. London: Heinemann.

Disse, M. and H. Engel (2001) 'Flood events in the Rhine Basin: Genesis, influences and mitigation', *Natural Hazards*, 23(2–3), 271–290.

Down, A. (1978) *Chichester Excavations 3*. Chichester: Phillimore.

Down, A. (1988) *Roman Chichester*. Chichester: Phillimore.

Down, A. and J. Bayley (1978) *Chichester Excavations 3*. Chichester: Phillimore.

Drake, B. L. (2017) 'Changes in North Atlantic oscillation drove population migrations and the collapse of the western Roman Empire', *Scientific Reports*, 7(1), 1227.

Draper, J. and C. Chaplin (1982) *Dorchester Excavations*. Dorchester: Dorset Natural History and Archaeological Society.

Drury, P. J. (1988) *The Mansio and Other Sites in the South-eastern Sector of Caesaromagus*. London: Chelmsford Archaeological Trust and Council for British Archaeology.

Duncan-Jones, R. (1990) *Structure and Scale in the Roman Economy*. Cambridge: Cambridge University Press.

Durrani, N. (2004) 'Tabard Square excavations, Southwark', *Current Archaeology*, 192, 540–547.

Dyson, T. (1978) 'Two Saxon land grants from Queenhithe', in J. Bird, H. Chapman and J. Clark (eds) *Collectanea Londiniensia: Studies in London Archaeology and History Presented to Ralph Merrifield*. London: LAMAS, 200–215.

Edgeworth, M. (2011) *Fluid Pasts: Archaeology of Flow*. London: Bristol Classical Press.

Edlund-Berry, I. (2006) 'Hot, cold, or smelly: The power of sacred water in Roman religion, 400–100 BCE', in C. E. Schultz and P. Harvey (eds) *Religion in Republican Italy*. Cambridge: Cambridge University Press, 162–180.

Eckhardt, H. (2006) 'The character, chronology, and use of the Late Roman pits: The Silchester finds assemblage', in M. Fulford, A. Clarke and H. Eckhardt (eds) *Life and Labour in Late Roman Silchester. Excavations in Insula IX since 1997*, Britannia Monograph Series No. 22. London: Society for Promotion of Roman Studies.

Ellis, P. (2000) *The Roman Baths and Macellum at Wroxeter: Excavations by Graham Webster 1955–85*. London: English Heritage.

Ennius (trans. E. H. Warmington, 1935) *Annales*. Cambridge, MA: Harvard University Press.

Esmonde Cleary, A. S. (2005) 'Beating the bounds: Ritual and the articulation of urban space in Roman Britain', in A. Mac Mahon and J. Price (eds) *Roman Working Lives and Urban Living*. Oxford: Oxbow Books, 1–17.

Esmonde Cleary, A. S. (2015) 'Review of temples and suburbs: Excavations at Tabard Square, Southwark', *Transactions of London and Middlesex Archaeological Society*, 66, 309–310.

Evans, C. (1997) 'Hydraulic communities: Iron Age enclosure in the East Anglian fenlands', in A. Gwilt and C. Haselgrove (eds) *Reconstructing Iron Age Societies: New Approaches to the British Iron Age*. Oxford: Oxbow Books, 216–227.

Fabbricoti, E. (1976) 'I bagni nelle prime ville romane', *Cronache Pompeianne*, 2, 29–111.

Fagan, B. M. (2012) *Elixir: A History of Water and Humankind*. New York: Bloomsbury Press.

Fagan, G. G. (1996) 'Sergius Orata: Inventor of the hypocaust?', *Phoenix*, 50(1), 56.

Fagan, G. G. (1999a) *Bathing in Public in the Roman World*. Ann Arbor, MI: University of Michigan Press.

Fagan, G. G. (1999b) 'Did slaves bathe at the baths?', in J. DeLaine and D. Johnston (eds) *Roman Baths and Bathing: Proceedings of the First Internation Conference on Roman Baths Held at Bath, England, 30 March–4 April 1992*. Portsmouth, RI: Journal of Roman Archaeology Supplementary Series 37, 25–34.

Fagan, G. G. (2001) 'The genesis of the Roman public bath: Recent approaches and future directions', *American Journal of Archaeology*, 105(3), 403–426.

Fagan, G. G. (2006) 'Bathing for health with Celsus and Pliny the Elder', *Classical Quarterly*, 56(1), 190–207.

Fairman, A. (2013) *Thameslink Archaeological Assessment 2: Excavations at 11–15 Borough High Street*. London: Pre-Construct Archaeology.

Farley, J. (2011) 'The deposition of miniature weaponry in Iron Age Lincolnshire', *Pallas*, 86, 97–121.

Ferraby, R. (2017) Personal communication.

Ferris, I. (2002) *Romano-British Religious Sites in the West Midlands*. Birmingham: West Midlands Regional Research Framework.

Ferris, I. (2007) 'Romano-British religious sites in the West Midlands, Birmingham', in R. Hingley and S. Willis (eds) *Roman Finds: Context and Theory. Proceedings of a Conference Held at the University of Durham, July 2002*. Oxford: Oxbow Books, 116–127.

Field, N. and M. Parker-Pearson (2003) *Fiskerton: An Iron Age Timber Causeway with Iron Age and Roman Votive Offerings: The 1981 Excavations*. Oxford: Oxbow Books.

Finley, M. I. (1973) *The Ancient Economy*. Berkeley, CA: University of California Press.

Fitzpatrick, A. P. (1984) 'The deposition of La Tene Iron Age metalwork in watery contexts in southern England', in B. Cunliffe and D. Miles (eds) *Aspects of the Iron Age in Central Southern Britain*. Oxford: Oxford University Committee for Archaeology, Institute of Archaeology, 178–190.

Ford, B. and S. Teague (2011) *Winchester – A City in the Making: Archaeological Excavations Between 2002 and 2007 on the Sites of Northgate House, Staple Gardens and the Former Winchester Library, Jewry St*. Oxford: Oxford Archaeology.

Fox, A. (1948) 'The early plan and town houses of Silchester (Calleva Atrebatum)', *Antiquity*, 22(88), 172–178.

Fox, G. E. and W. H. S. J. Hope (1894) 'Excavations on the site of the Roman city at Silchester, in 1893', *Archaeologia*, 54(1), 199–238.

Fox, G. E. and W. H. S. J. Hope (1901) 'Excavations on the site of the Roman city at Silchester, in 1900', *Archaeologia*, 57(2), 229–256.

Fraser, C. (2009) *Rewilding the World: Dispatches from the Conservation Revolution*. New York: Metropolitan Books.

Frere, S. S. (1947) *Roman Canterbury; The City of Durovernum*. London: Medici Society.

Frere, S. S. (1967) *Britannia: A History of Roman Britain*. London: Routledge & Keegan Paul.

Frere, S. S. (1972) *Verulamium Excavations I*. Oxford: Society of Antiquaries of London.

Frere, S. S. and R. S. O. Tomlin (1991) 'Roman Britain in 1990', *Britannia*, 22, 222–311.

Frere, S. S. and M. G. Wilson (1983) *Verulamium Excavations II*. London: Society of Antiquaries of London.

Frere, S. S., M. W. C. Hassall and R. S. O. Tomlin (1984) 'Roman Britain in 1983', *Britannia*, 15, 266–356.

Frere, S. S., M. W. C. Hassall and R. S. O. Tomlin (1988) 'Roman Britain in 1987', *Britannia*, 19, 415–508.

Frere, S. S., M. W. C. Hassall and R. S. O. Tomlin (1990) 'Roman Britain in 1989', *Britannia*, 21, 303–378.

Frontinus (trans. M. B. McElwain, 1925) *De aquaeductu urbis (On the Aqueducts of Rome)*. London: Heinemann.

Fulford, M. G. (2001) 'Links with the past: Pervasive "ritual" behaviour in Roman Britain', *Britannia*, 32, 199–218.

Fulford, M. G. and J. Timby (2000) *Late Iron Age and Roman Silchester: Excavations on the Site of the Forum-basilica, 1977, 1980–86*. London: Society for the Promotion of Roman Studies.

Gaffney, V. L., R. H. White and H. Goodchild (2007) *Wroxeter, the 'Cornovii', and the Urban Process: Final Report on the Wroxeter Hinterland Project 1994–1997*, Vol. 1. Portsmouth, RI: Journal of Roman Archaeology.

Galen (trans. R. M. Green, 1951) *De sanitate tuenda (On Hygiene)*. Springfield, IL: Charles C. Thomas.

García Quintela, M. V. and M. Santos-Estévez (2015) 'Iron Age saunas of northern Portugal: State of the art and research perspectives', *Oxford Journal of Archaeology*, 34(1), 67–95.

Garland, N. (2017) *Territorial Oppida and the Transformation of Landscape and Society in South-eastern Britain from BC300 to 100 AD*. London: University College London.

Gerard, J. (2011) 'Wells and belief systems at the end of Roman Britain: A case study from Roman London', in L. Lavan and M. Mulryan (eds) *The Archaeology of Late Antique 'Paganism'*. Leiden: Brill, 551–574.

Going, C. J. and J. R. Hunn (1999) *Excavations at Boxfield Farm, Chells, Stevenage, Hertfordshire. Report No. 2*. Hertford: Hertfordshire Archaeological Trust.

Goodburn, R., M. W. C. Hassall and R. S. O. Tomlin (1979) 'Roman Britain in 1978', *Britannia*, 10, 268–356.

Goodman, P. J. (2007) *The Roman City and Its Periphery: From Rome to Gaul*. London: Routledge.

Goodyear, F. R. (1983) 'Technical writing', in W. Clausen (ed.) *The Cambridge History of Classical Literature, II*. Cambridge: Cambridge University Press.

Green, M. (1989) *Symbol and Image in Celtic Religious Art*. London: Routledge.

Greene, M. J. (2015) *Beyond the Hinterland: Understanding Late Iron Age Transitional Rural Settlement and Society in South Shropshire*. Birmingham: University of Birmingham.

Grünthal, G., A. H. Thieken, J. Schwarz, K. S. Radtke, A. Smolka and B. Merz (2006) 'Comparative risk assessments for the city of Cologne – Storms, floods, earthquakes', *Natural Hazards*, 38(1/2), 21–44.

Gunner, W. (1849) 'A Roman watercourse in Winchester', *Archaeological Journal*, 6, 408.

Hanson, J. (1971) 'Municipal and military water supply and drainage in Roman Britain', PhD thesis, University of London.

Hartnett, J. (2008) 'Fountains at Herculaneum: Sacred history, topography, and civic identity', *Rivista di Studi Pompeiani*, 19, 77–89.

Haselgrove, C. (1989) 'The later Iron Age in southern Britain and beyond', in M. Todd and T. F. C. Blagg (eds) *Research on Roman Britain, 1960–89*. London: Society for Promotion of Roman Studies, 1–18.

Haverfield, F. (1912) *The Romanization of Roman Britain*. Oxford: Clarendon Press.

Haverfield, F. (1913) *Ancient Town-Planning*. Oxford: Clarendon.

Hawkes, C. F. C. and P. Crummy (1995) *Camulodunum 2*. Colchester: Colchester Archaeological Trust.

Hawkes, C. F. C. and M. Hull (1947) *Camulodunum: First Report on the Excavations at Colchester, 1930–1939*. Oxford: Oxford University Press.

Heard, K., H. Sheldon and P. Thompson (1990) 'Mapping Roman Southwark', *Antiquity*, 64(244), 608–619.

Henderson, C. (1988) 'Exeter (Isca Dumnoniorum)', in G. Webster (ed.) *Fortress into City: The Consolidation of Roman Britain First Century AD*. London: Batsford, 91–120.

Henig, M. (1984) *Religion in Roman Britain*. London: Batsford.

Henig, M. (1995) *The Art of Roman Britain*. London: Batsford.

Henig, M. (1998) 'The temple as a bacchium or sacrarium in the fourth century', in J. D. Shepherd (ed.) *The Temple of Mithras, London: Excavations by W. F. Grimes and A. Williams at the Walbrook*. London: English Heritage, 230–235.

Henig, M. and G. Webster (2002) 'The Venus (or nymph) of the fountain', in J. Chadderton and G. Webster (eds) *The Legionary Fortress at Wroxeter: Excavations by Graham Webster, 1955–1985*. London: English Heritage, 135–136.

Herget, J. and H. Meurs (2010) 'Reconstructing peak discharges for historic flood levels in the city of Cologne, Germany', *Global and Planetary Change*, 70(1), 108.

Hill, J. D. (1995) *Ritual and Rubbish in the Iron Age of Wessex: A Study on the Formation of a Specific Archaeological Record*. Oxford: British Archaeological Reports.

Hill, J. D. (2011) 'How did British middle and late pre-Roman Iron Age societies work (if they did)?', in T Moore and X. Armada (eds) *Atlantic Europe in the First Millenium BC*. Oxford: Oxford University Press, 242–263.

Hill, J. D. and P. Rowsome (2011) *Roman London and the Walbrook Stream Crossing: Excavations at 1 Poultry and Vicinity, City of London*. London: Museum of London Archaeology.

Hingley, R. (1982) 'Recent discoveries of the Roman period at the Noah's Ark Inn, Frilford, South Oxfordshire', *Britannia*, 13, 305–309.

Hingley, R. (2000) *Roman Officers and English Gentlemen: The Imperial Origins of Roman Archaeology*. London: Routledge.

Hingley, R. (2001) 'Images of Rome', in R. Hingley (ed.) *Images of Rome: Perceptions of Ancient Rome in Europe and the United States in the Modern Age*. Portsmouth, RI: Journal of Roman Archaeology, 7–22.

Hingley, R. (2005) *Globalizing Roman Culture: Unity, Diversity and Empire*. London: Routledge.

Hingley, R. (2006) 'The deposition of iron objects in Britain during the later prehistoric and Roman periods: Contextual analysis and the significance of iron', *Britannia*, 37, 213–257.

Hingley, R. (2012) *Hadrian's Wall: A Life*. Oxford: Oxford University Press.

Hodder, I. and M. Hassall (1971) 'The non-random spacing of Romano-British walled towns', *Man*, 6, 391–407.

Hodge, T. (1992) *Roman Aqueducts & Water Supply*. London: Duckworth.

Holbrook, N. (1998) 'Shops V.1–V.5 in Insula V: Excavations directed by J. S. Wacher', in N. Holbrook (ed.) *Cirencester: The Roman Town Defences, Public Buildings and Shops*. Cirencester: Cotswold Archaeological Trust, 189–209.

Holbrook, N. and J. P. Salvatore (1998) 'The street system', in N. Holbrook (ed.) *Cirencester: The Roman Town Defences, Public Buildings and Shops*. Cirencester: Cotswold Archaeological Trust, 19–31.

Holland, L. A. (1961) *Janus and the Bridge*. Rome: American Academy in Rome.

Hooker, E. (1963) 'The significance of Numa's religious reforms', *Numen*, 10(2), 87–132.

Hope, W. H. (1897) 'Excavations on the site of the Roman city at Silchester, Hants, in 1896', *Archaeologia*, 55(2), 409–430.

Hopkins, J. (2007) 'The Cloaca Maxima and the monumental manipulation of water in archaic Rome', *Waters of Rome*, 4, 1–15.

Hopkins, J. (2012) 'The "sacred sewer": Tradition and religion in the Cloaca Maxima', in M. Bradley (ed.) *Rome, Pollution and Propriety: Dirt, Disease and Hygiene in the Eternal City from Antiquity to Modernity*. Cambridge: Cambridge University Press, 81–102.

Hull, M. D. (1958) *Roman Colchester*. Oxford: Society of Antiquaries.

Hunn, J., D. S. Neal and A. Wardle (1990) *Excavation of the Iron Age, Roman and Medieval Settlement at Gorhambury, St. Albans*. London: English Heritage.

Hurst, H. (1988) 'Gloucester (Glevum)', in G. Webster (ed.) *Fortress into City: The Consolidation of Roman Britain First Century* AD. London: Batsford, 48–74.

Isserlin, R. M. (1998) 'A spirit of improvement? Marble and the culture of Roman Britain', in R. Laurence and J. Berry (eds) *Cultural Identity in the Roman Empire*. London: Routledge, 125–156.

Izdebski, A., K. Holmgren, E. Weiberg, S. R. Stocker, U. Büntgen, A. Florenzano, A. Gogou, S. A. G. Leroy, J. Luterbacher, B. Martrat, A. Masi, A. M. Mercuri, P. Montagna, L. Sadori, A. Schneider, M.-A. Sicre, M. Triantaphyllou and E. Xoplaki (2016) 'Realising consilience: How better communication between archaeologists, historians and natural scientists can transform the study of past climate change in the Mediterranean', *Quaternary Science Reviews*, Special Issue: Mediterranean Holocene Climate, Environment and Human Societies, 136, 5–22.

Jackson, R. (1990) 'Waters and spas in the classical world', *Medical History*, 10, 1–13.

Jackson, R. (1999) 'Spas, waters and hydrotherapy in the Roman world', in J. DeLaine and D. Johnston (eds) *Roman Baths and Bathing: Proceedings of the First International Conference on Roman Baths Held at Bath, England, 30 March–4 April 1992*. Portsmouth, RI: Journal of Roman Archaeology Supplementary Series 37, 107–116.

James, H. (1993) 'Roman Carmarthen', in S. J. Greep (ed.) *Roman Towns: The Wheeler Inheritance; A Review of 50 Years' Research*. York: Council for British Archaeology, 93–98.

Jarman, C. (1997) 'Christ Church College', in *Canterbury's Archaeology 1995–1996*. Canterbury: Canterbury Archaeological Trust, 2–5.

Johnson, A. (1983) *Roman Forts of the 1st and 2nd Centuries* AD *in Britain and the German Provinces*. London: Adam & Charles Black.

Jones, A. H. M. (1974) *The Roman Economy: Studies in Ancient Economic and Administrative History*. Oxford: Blackwell.

Jones, F. (1992) *The Holy Wells of Wales*. Cardiff: University of Wales Press.

Jones, M. J. (2003) 'Sources of effluence: Water through Roman Lincoln', in J. Wacher and P. R. Wilson (eds) *The Archaeology of Roman Towns: Studies in Honour of John S. Wacher*. Oxford: Oxbow Books, 111–128.

Jones, M. J., D. Stocker and A. Vince (2003a) *The City by the Pool: Assessing the Archaeology of the City of Lincoln*. Oxford: Oxbow Books.

Jones, M. J., D. Stocker, A. G. Vince, D. Powlesland and J. Herridge (2003b) 'Lincoln archaeological research assessment research agenda zones, GIS and interactive database', available at www.heritageconnectlincoln.com/.

Joyce, J. G. (1881) 'XVI – Third account of excavations at Silchester', *Archaeologia Archaeologia*, 46(2), 344–365.

Kaika, M. (2005) *City of Flows: Modernity, Nature, and the City*. New York: Routledge.

Kaika, M. and E. Swyngedouw (2000) 'Fetishizing the modern city: The phantasmagoria of urban technological networks', *International Journal of Urban and Regional Research*, 24(1), 120–138.

Kaiser, A. (2000) *The Urban Dialogue: An Analysis of the Use of Space in the Roman City of Empúries, Spain*. Oxford: Archaeopress.

Kamash, Z. (2012) 'An exploration of the relationship between shifting power, changing behavior and new water technologies in the Roman Near East', *Water History*, 4(1), 79–93.

Keen, L. (1977) 'Dorset archaeology in 1977', *Dorset Natural History and Archaeology Society*, 99, 120–126.

Kenyon, K. M. (1948) *Excavations at the Jewry Wall Site, Leicester*. London: Society of Antiquaries.

Killock, D., J. D. Shepherd, J. Gerard, K. Hayward, K. Rielly and V. Ridgeway (2015) *Temples and Suburbs: Excavations at Tabard Square, Southwark*. London: Pre-Construct Archaeology.

King, A. (2001) 'Gallic symbols in French politics and culture', in R. Hingley (ed.) *Images of Rome: Perceptions of Ancient Rome in Europe and the United States in the Modern Age*. Portsmouth, RI: Journal of Roman Archaeology, 113–125.

King, A. (2005) 'Animal remains from temples in Roman Britain', *Britannia*, 36, 329.

King, A. and G. Soffe (1983) 'A Romano-Celtic temple at Ratham Mill, Funtington, West Sussex', *Britannia*, 14, 264–266.

King, C. and P. J. Woodward (2003) 'The 1986 Poundbury hoard of 3rd century Antoniniani', *Proceedings of the Dorset Archaeology and Natural History Society*, 125, 150–152.

Kipling, R. (1906) *Puck of Pook's Hill*. London: Macmillan.

Kipling, R. (1911) 'The Roman centurion's song', available at www.kiplingsociety.co.uk/poems_romancenturion.htm.

Knüsel, C. J. and G. C. Carr (1995) 'On the significance of the crania from the River Thames and its tributaries', *Antiquity*, 69(262), 162–169.

Lambot, B. (1998) 'Les Morts d'Acy-Romance (Ardennes) à LaTène finale: Pratiques funéraires, aspects religieuses et hiérarchie sociale', in *Les Celtes: Rites funéraires en Gaule du Nord entre le 6 et le 1 siècle avant J.-C*. Namur: Ministère de la Région Wallonne, 75–87.

Lane, R. (2018) Personal communication.

Larsson, L. (2001) 'South Scandinavian wetland sites and finds from the Mesolithic and the Neolithic', in B. Purdy (ed.) *Enduring Records: The Environmental and Cultural Heritage of Wetlands*. Oxford: Oxbow Books, 158–171.

Laurence, R. (1994a) 'Modern ideology and the creation of ancient town planning', *European Review of History*, 1(1), 9–18.

Laurence, R. (1994b) *Roman Pompeii: Space and Society*. London: Routledge.

Laurence, R. (2010) *Roman Passions: A History of Pleasure in Imperial Rome.* London: Continuum.

Laurence, R., A. S. Esmonde Cleary and G. Sears (2011) *The City in the Roman West, c.250 BC–c. AD 250.* Cambridge: Cambridge University Press.

Laurie, T. C. (2003) 'Researching the prehistory of Wensleydale, Swaledale and Teesdale', in T. G. Manby, S. Moorhouse and P. Ottoway (eds) *The Archaeology of Yorkshire: An Assessment at the Beginning of the 21st Century.* Leeds: Yorkshire Archaeological Society, 223–254.

Lees, D. (1989) 'Excavations in the Walbrook Valley', *London Archaeologist,* 6, 115–119.

Leveau, P. (1991) 'Research on Roman aqueducts in the past ten years', in T. Hodge (ed.) *Future Currents in Aqueduct Studies.* Leeds: F. Cairns, 149–162.

Leveau, P. (1996) 'The Barbegal water mill in its environment: Archaeology and the economic and social history of antiquity', *Journal of Roman Archaeology,* 9, 137–153.

Lewis, M. J. T. (1984) 'Our debt to Roman engineering: The water supply of Lincoln to the present day', *Industrial Archaeology Review,* 7(1), 57–73.

Liversidge, J., D. J. Smith, I. M. Stead and V. Rigby (1973) 'Brantingham Roman Villa: Discoveries in 1962', *Britannia,* 4, 84–106.

Livy (trans. C. Roberts, 1905) *Ab Urbe Condita (History of Rome).* London: J. M. Dent and Sons.

Ljungqvist, F. C. (2010) 'A new reconstruction of temperature variability in the extra-tropical northern hemisphere during the last two millenia', *Geografiska Annaler Series A: Physical Geography,* 92(3), 339–351.

Lloris, F. B. (2006) 'An irrigation decree from Roman Spain: "The Lex Rivi Hiberiensis"', *Journal of Roman Studies,* 96, 147–197.

Lodwick, L. (2015) 'An archaeobotanical analysis of Silchester and the wider region across the Late Iron Age–Roman transition', PhD thesis, University of Oxford.

Lucas, C. P. (1912) *Greater Rome and Greater Britain.* Oxford: Clarendon Press.

Lyne, M. (1999) 'The pottery from the lower slope', in R. Niblett (ed.) *The Excavation of a Ceremonial Site at Folly Lane, Verulamium,* Britannia Monograph. London: Society for Promotion of Roman Studies, 233–307.

Macfarlane, R. (2015) *Landmarks.* London: Hamish Hamilton.

MacGregor, A. (1976) *Finds from a Roman Sewer System and an Adjacent Building in Church Street.* London: York Archaeological Trust/Council for British Archaeology.

MacGregor, A. (1978) *Roman Finds from Skeldergate and Bishophill.* London: York Archaeological Trust/Council for British Archaeology.

Mackie, N. (1990) 'Urban munificence and the growth of urban consciousness', in T. F. C. Blagg and M. Millett (eds) *The Early Roman Empire in the West.* Oxford: Oxbow Books, 179–193.

Macnaghten, P. and J. Urry (1998) *Contested Natures.* London: Sage.

Magilton, J. (1996) 'Chichester, the Burghal Hidage and the diversion of the River Lavant', *Archaeology of Chichester and District 1996.* Chichester: Chichester District Council, 37–41.

Malacrino, C. G. (2010) *Constructing the Ancient World Architectural Techniques of the Greeks and Romans.* San Giovanni Lupatoto: Arsenale Editrice.

Manning, W. H. (1985) *Catalogue of the Romano-British Iron Tools, Fittings and Weapons in the British Museum.* London: British Museum Publications.

Maréchal, S. (2013) 'Roman public baths in modern Libya', *BABESCH*, 88, 205–228.

Marsden, P. (1975) 'The excavation of a Roman palace site in London 1961–72', *Transactions of the London and Middlesex Archaeological Society*, 26, 36–41.

Marsden, P. (1987) *The Roman Forum Site in London: Discoveries before 1985.* London: Museum of London.

Marsh, G. and B. West (1981) 'Skullduggery in Roman London?', *Transactions of the London and Middlesex Archaeological Society*, 32, 86–102.

Martial (trans. W. C. A. Ker, 1920) *Epigrammata (Epigrams).* London: Heinemann.

Mattingly, D. J. (2007) *An Imperial Possession: Britain in the Roman Empire, 54 BC–AD 409.* London: Penguin.

Mattingly, D. J. (2008) 'Urbanism, epigraphy and identity in the towns of Britain under Roman rule', in A. Birley, H. M. Schellenberg, V. E. Hirschmann and A. Krieckhaus (eds) *A Roman Miscellany: Essays in Honour of Anthony R. Birley on His Seventieth Birthday.* Gdańsk: Foundation for Development of Gdańsk University, 53–71.

May, J. (1976) *Prehistoric Lincolnshire.* Lincoln: History of Lincolnshire Committee.

May, J. (1988) 'Iron Age Lincoln? The topographical and settlement evidence reviewed', *Britannia*, 19, 50–57.

Merrifield, R. (1983) *London, City of the Romans.* London: Batsford.

Merrifield, R. (1987) *The Archaeology of Ritual and Magic.* London: Batsford.

Merrifield, R. and J. Hall (2008) 'In its depths what treasures – The nature of the Walbrook stream valley and the Roman metalwork found therein', in J. Clark, J. Cotton, J. Hall, R. Sherris and H. Swain (eds) *Londinium and Beyond: Essays on Roman London and its Hinterland for Harvey Sheldon.* York: Council for British Archaeology, 121–127.

Miller, L., J. Schofield and M. Rhodes (1986) *The Roman Quay at St Magnus House, London: Excavations at New Fresh Wharf, Lower Thames Street, London 1974–78.* London: Museum of London.

Millett, M. (1990) *The Romanization of Britain: An Essay in Archaeological Interpretation.* Cambridge: Cambridge University Press.

Millett, M. (2001) 'Approaches to urban societies', in S. James and M. Millett (eds) *Britons and Romans: Advancing an Archaeological Agenda.* York: Council for British Archaeology, 60–67.

Millett, M. (2006) 'Characterizing Shiptonthope', in M. Millett and L. Allason-Jones (eds) *Shiptonthorpe, East Yorkshire: Archaeological Studies of a Romano-British Roadside Settlement.* Leeds: Yorkshire Archaeological Society, Roman Antiquities Section and East Riding Archaeological Society, 305–318.

Millett, M. (2017) Personal communication.

Milne, G. (1985) *The Port of Roman London.* London: Batsford.

Milne, G. (1992) *From Roman Basilica to Medieval Market: Archaeology in Action in the City of London.* London: HMSO.

Milne, G. (1996) 'A palace disproved: Reassessing the provincial governor's presence in 1st century London', in J. Bird, M. Hassall and H. Sheldon (eds)

Interpreting Roman London: Papers in Memory of Hugh Chapman. Oxford: Oxbow Books, 49–55.

Mommsen, T. (1885) *Römische Geschichte*. Leipzig: Reimer & Hirsel.

Monbiot, G. (2014) *Feral: Rewilding the Land, the Sea, and Human Life*. Chicago, IL: University of Chicago Press.

Morley, N. (2004) *Theories, Models, and Concepts in Ancient History*. New York: Routledge.

Morton, O. (2016) *The Planet Remade: How Geoengineering Could Change the World*. Princeton, NJ: Princeton University Press.

Mount, F. (2010) *Full Circle: How the Classical World Came Back to Us*. London: Simon & Schuster.

Murphy, T. M. (2004) *Pliny the Elder's Natural History: The Empire in the Encyclopedia*. Oxford: Oxford University Press.

Murray, M. (1995) 'Viereckschanzen and feasting: Socio-political ritual in Iron-Age Central Europe', *Journal of European Archaeology*, 3(2), 125–151.

Myers, S. (2011) *Walking on Water: London's Hidden Rivers Revealed*. Chalford: Amberley Publishing.

Neal, D. S. (1974) *The Excavation of the Roman Villa in Gadebridge Park, Hemel Hempstead, 1963–8*. London: Society of Antiquaries of London.

Niblett, R. (1999) *The Excavation of a Ceremonial Site at Folly Lane, Verulamium*. London: Society for Promotion of Roman Studies.

Niblett, R. (2001) *Verulamium: The Roman City of St Albans*. Stroud: Tempus.

Niblett, R. (2005) 'Roman Verulamium', in R. Niblett and I. Thompson (eds) *Alban's Buried Towns: An Assessment of St. Albans' Archaeology up to AD 1600*. Oxford: Oxbow Books/English Heritage, 41–166.

Niblett, R. (1985) *Sheepen: An Early Roman Industrial Site at Camulodunum*. London: Council for British Archaeology.

Niblett, R. and I. Thompson (eds) (2005) *Alban's Buried Towns: An Assessment of St. Albans' Archaeology up to AD 1600*. Oxford: Oxbow Books/English Heritage.

Nielsen, I. (1993) *Thermae et Balnea: The Architecture and Cultural History of Roman Public Baths*. Aarhus: Aarhus University Press.

Nriagu, J. O. (1983) *Lead and Lead Poisoning in Antiquity*. New York: Wiley.

Oleson, J. P. (1984) *Greek and Roman Mechanical Water-lifting Devices: The History of a Technology*. Toronto, ON: University of Toronto Press.

Oliver, M. (1992) 'Excavation of an Iron Age and Romano-British settlement site at Oakridge, Basingstoke, Hampshire, 1965–6', *Proceedings of Hampshire Field Club and Archaeological Society*, 48, 55–94.

Orlin, E. M. (2002) *Temples, Religion, and Politics in the Roman Republic*. Leiden: Brill.

O'Sullivan, P. (2008) 'The "collapse" of civilizations: What palaeoenvironmental reconstruction cannot tell us, but anthropology can', *The Holocene*, 18(1), 45–55.

Ottaway, P. (1993) *English Heritage Book of Roman York*. London: Batsford.

Ottaway, P. (2010) *Heslington East, York: A Second Addendum to the Archaeological Remains Management Plan*. York: PJO Archaeology.

Ovid (trans. G. Showerman, 1931) *Amores*. London: Heinemann.

Ovid (trans. G. Showerman, 1931) *Heroides*. London: Heinemann.

Ovid (trans. J. G. Frazer, 1931) *Fasti*. London: Heinemann.

Ovid (trans. A. D. Melville, 1986) *Metamorphoses*. Oxford: Oxford University Press.

Papworth, M. (2011) *The Search for the Durotriges: Dorset and the West Country in the Late Iron Age*. Stroud: History Press.

Parcero Oubiña, C. and I. Cobas Fernández (2004) 'Iron Age archaeology of the northwest Iberian Peninsula', *e-Keltoi: Journal of Interdisciplinary Celtic Studies*, 6, 1–72.

Perring, D. (1991) *Roman London*. London: Seaby.

Perry, B. (1966) 'Some recent discoveries in Hampshire', in A. Thomas (ed.) *Rural Settlement in Roman Britain: Papers Given at a C.B.A. Conference Held at St. Hugh's College Oxford, January 1 to 3, 1965*. London: Council for British Archaeology, 39–42.

Petch, D. (1962) 'A Roman inscription, Nettleham', *Lincolnshire Architecture and Archaeological Society Reports*, 9(2), 94–97.

Phillips, T. (2004) 'Seascapes and landscapes in Orkney and northern Scotland', *World Archaeology*, 35(3), 371–384.

Pliny the Elder (trans. H. Rackham, 1942) *Historia Naturalis (Natural History)*. London: Heinemann.

Pliny the Younger (trans. B. Radice, 1963) *Epistulae (The Letters of the Younger Pliny)*. London: Penguin.

Plutarch (trans. B. Perrin, 1923) *Vitae Romulus (Life of Romulus)*. Cambridge, MA: Harvard University Press.

Plutarch (trans. I. Scott-Kilvert, 1965) *Vitae Coriolanus (Life of Coriolanus)*. London: Penguin.

Pollard, R. (1996) 'Iron Age riverside pit alignments at St. Ives, Cambridgeshire', *Proceedings of the Prehistoric Society*, 62, 93–116.

Pollard, R. (1998) 'A ceramic cult figure from Leicester', *Britannia*, 29, 353–356.

Potter, T. W. and C. Johns (1992) *Roman Britain*. Berkeley, CA: University of California Press.

Poux, M. (2000) 'Espaces votifs – Espaces festifs. Banquets et rites de libation en contexte de sanctuaires et d'enclose', *Revue archéologique de Picardie*, 1(1), 217–231.

Pratt, S. (2009) 'Two "new" town gates, Roman buildings and an Anglo-Saxon sanctuary at St. Mildred's Tannery, Canterbury', *Archaeologia Cantiana*, CXXIX, 225–238.

Pryor, F. (1992) 'Discussion: The Fengate/Northey landscape', *Antiquity*, 66(251), 518–531.

Pucci, S., D. Pantosti, P. M. De Martini, A. Smedile, M. Munzi, E. Cirelli, M. Pentiricci and L. Musso (2011) 'Environment–human relationships in historical times: The balance between urban development and natural forces at Leptis Magna (Libya)'. *Quaternary International*, 242, 171–184.

Purcell, N. (1990) 'The creation of a provincial landscape: The Roman impact on Cisalpine Gaul', in T. F. C. Blagg and M. Millett (eds) *The Early Roman Empire in the West*. Oxford: Oxbow Books, 7–29.

Purcell, N. (1996) 'Rome and the management of water: Environment, culture and power', in G. Shipley and J. B. Salmon (eds) *Human Landscapes in Classical Antiquity: Environment and Culture*. London: Routledge, 180–212.

Putnam, B. (1997) 'The Dorchester Roman aqueduct', *Current Archaeology*, 154, 364–369.

Quinn, P. (1999) *The Holy Wells of Bath and Bristol Region*. Woonton Almeley: Logaston Press.

Ralph, S. (2007) *Feasting and Social Complexity in Later Iron Age East Anglia*. Oxford: Archaeopress.

Rattue, J. (1995) *The Living Stream: The Holy Wells in Historical Context*. Woodbridge: Boydell.

Reece, R. (2003) 'The siting of Roman Corinium', *Britannia*, 34, 276–280.

Reece, R. and P. Broxton (2011) 'Cirencester: The interpretation of streams', *Cirencester Archaeological and Historical Society*, Vol. 53. Cirencester: Cirencester Archaeological and Historical Society.

Rhodes, M. (1991) 'The Roman coinage from London Bridge and the development of the city and Southwark', *Britannia*, 22, 179–190.

RIB (1985–) *The Roman Inscriptions of Britain*. Oxford: Clarendon Press.

Rivet, A. L. F. (1958) *Town and Country in Roman Britain*. London: Hutchinson University Library.

Rivet, A. L. F. and C. Smith (1979) *The Place-Names of Roman Britain*. Princeton, NJ: Princeton University Press.

Robinson, D. N. (2009) 'Storms of the wolds', in D. N. Robinson (ed.) *The Lincolnshire Wolds*. Oxford: Windgather Press, 75–76.

Rodgers, R. H. (2003) *Introduction: Frontinus – De Aquaeductu Urbis Romae*. Cambridge: Cambridge University Press.

Rogers, A. (2008) 'Roman towns as meaning-laden places reconceptualising the growth and decline of towns in Roman Britain, Volume 1', PhD thesis, University of Durham.

Rogers, A. (2011a) *Late Roman Towns in Britain: Rethinking Change and Decline*. New York: Cambridge University Press.

Rogers, A. (2011b) 'Reimagining Roman ports and harbours: The port of Roman London and waterfront archaeology', *Oxford Journal of Archaeology*, 30(2), 207–225.

Rogers, A. (2013) *Water and Roman Urbanism: Towns, Waterscapes, Land Transformation and Experience in Roman Britain*. Leiden: Brill.

Rogers, A. and R. Hingley (2010) 'Edward Gibbon and Francis Haverfield: The traditions of imperial decline', in M. Bradley (ed.) *Classics and Imperialism in the British Empire*. Oxford: Oxford University Press, 189–209.

Rook, T. (1992) *Roman Baths in Britain*. Princes Risborough: Shire Publications.

Ross, A. (1968) 'Shafts, pits, wells – Sanctuaries of the Belgic Britons', in J. M. Coles and D. D. A. Simpson (eds) *Studies in Ancient Europe: Essays Presented to Stuart Piggott*. Leicester: Leicester University Press, 255–285.

Rostovtzeff, M. I. (1926) *The Social and Economic History of the Roman Empire*. Oxford: Clarendon Press.

Rowsome, P. (1998) 'The development of the town plan of early Roman London', *Journal of Roman Archaeology*, Supplementary Series 24, 35–46.

Rowsome, P. (1999) 'The Huggin Hill Baths and bathing in London', in J. DeLaine and D. Johnston (eds) *Roman Baths and Bathing: Proceedings of the First International Conference on Roman Baths Held at Bath, England, 30 March–4 April 1992. Portsmouth, RI: Journal of Roman Archaeology Supplementary Series 37, 262–277.

Rowsome, P. (2008) 'Mapping Roman London: Identifying its urban patterns and interpreting their meaning', in J. Clark, J. Cotton, J. Hall, R. Sherris and H. Swain (eds) *Londinium and Beyond. Essays on Roman London and its Hinterland for Harvey Sheldon*. York: Council for British Archaeology, 25–32.

Rowsome, P., M. Burch, C. Lemos, T. Wellman, A. Chopping and S. Hirst (2011) *Londinium: A New Map and Guide to Roman London*. London: Museum of London Archaeology.

Russell, M., P. Cheetham, I. Hewitt, E. Hambleton, H. Manley and D. Stewart (2017) 'The Durotriges Project 2016: An interim statement', *Proceedings of the Dorset Natural History and Archaeology Society*, 138, 106–112.

Rykwert, J. (1976) *The Idea of a Town: Anthropology of Urban Form in Rome, Italy and the Ancient World*. London: Faber and Faber.

Rylatt, J. and B. Bevan (2007) 'Realigning the world: Pit alignments and their landscape context', in C. Haselgrove and T. Moore (eds) *The Later Iron Age in Britain and Beyond*. Oxford: Oxbow Books, 219–234.

Sauer, E. (2011) 'Religious rituals at springs', in L. Lavan and M. Mulryan (eds) *The Archaeology of Late Antique 'Paganism'*. Leiden: Brill, 505–550.

Saunders, C. (1974) 'Roman bath house, Branch Road, St Albans', *Hertfordshire Archaeological Review*, 9, 166–168.

Saunders, C. (1975) 'Notes on excavations at the Six Bells and the Branch Road bath house in 1974', *Britannia*, 6, 258–260.

Scarpino, P. V. (1997) *Large Floodplain Rivers as Human Artifacts: A Historical Perspective on Ecological Integrity*. Fort Belvoir, VA: Defense Technical Information Center.

Schumaker, L. (2008) 'Slimes and death-dealing dambos: Water, industry and the garden city on Zambia's Copperbelt', *Journal of Southern African Studies*, 34(4), 823–840.

Scullard, H. H. (1981) *Festivals and Ceremonies of the Roman Republic*. London: Thames & Hudson.

Sedlak, D. (2015) *Water 4.0: The Past, Present, and Future of the World's Most Vital Resource*. Reprint edition. New Haven, CT: Yale University Press.

Seeley, F. and A. Wardle (2009) 'Religion and ritual', in C. Cowan (ed.) *Roman Southwark Settlement and Economy: Excavations in Southwark, 1973–1991*. London: Museum of London Archaeology, 143–157.

Seneca (trans. J. Clarke, 1910) *Quaestiones naturales (Natural Questions)*. New York: Macmillan.

Servius (eds G. Thilo and H. Hagen, 1961) *In Vergilii Aeneidem commentarii (Commentary on Virgil)*. Hildesheim: Georg Olms.

Sharples, N. M. and J. Ambers (1991) *Maiden Castle: Excavations and Field Survey 1985–6*. London: English Heritage.

Shepherd, J. D. (1998) *The Temple of Mithras, London: Excavations by W. F. Grimes and A. Williams at the Walbrook*. London: English Heritage.

Shklovsky, V. (1965) 'Art as technique (1917)', in L. T. Lemon and M. J. Reis (trans.) *Russian Formalist Criticism: Four Essays*. Lincoln, NE: University of Nebraska Press, 3–24.

Smith, C. R. (1841) 'On the Roman coins discovered in the bed of the Thames near London Bridge from 1834 to 1841', *Numismatic Chronicle*, 4, 147–168.

Smith, J. R. (1922) *Springs and Wells in Greek and Roman Literature, Their Legends and Locations.* New York: G. P. Putnam & Sons.

Smith, T. (2009) 'Restoring biodiversity', in D. N. Robinson (ed.) *The Lincolnshire Wolds.* Oxford: Windgather Press, 103–114.

Southern, P. (2007) *The Roman Army: A Social and Institutional History.* Oxford: Oxford University Press.

Spain, R. (2004) 'A possible Roman tide mill', Paper 5, Kent Archaeological Society, Maidstone.

Sparey-Green, C. (2018) Personal communication.

Statius (trans. J. H. Mozley, 1928) *Silvae.* London: Heinemann.

Steane, K., M. J. Darling, M. Jones, J. Mann, A. Vince and J. Young (2006) *The Archaeology of the Upper City and Adjacent Suburbs.* Oxford: Oxbow Books.

Stephens, G. R. (1985a) 'Civic aqueducts in Britain', *Britannia*, 16, 197–208.

Stephens, G. R. (1985b) 'Military aqueducts in Roman Britain', *Archaeological Journal*, 142(1), 216–236.

Stirling, L. M., D. J. Mattingly and N. Ben Lazreg (2001) *Leptiminus (Lamta)/the East Baths, Cemeteries, Kilns, Venus Mosaic, Site Museum, and Other Studies.* Ann Arbor, MI: University of Michigan.

Stjernquist, B., T. Nilsson and U. Møhl (1997) *The Röekillorna Spring: Spring Cults in Scandinavian Prehistory.* Stockholm: Almqvist and Wiksell International.

Stocker, D. (1998) 'A hitherto unidentified image of the Mithraic god Arimanius at Lincoln?', *Britannia*, 29, 359–363.

Stocker, D. and P. Everson (2003) 'The straight and narrow way. Fenland causeways and the conversion of the landscape in the Witham Valley, Lincolnshire', in M. Carver (ed.) *The Cross Goes North. Processes of Conversion in Northern Europe, AD 300–1300.* York: York Medieval Press, 271–288.

Strabo (trans. H. L. Jones, 1917) *Geographica (Geography).* London: Heinemann.

Strang, V. (2004) *The Meaning of Water.* Oxford: Berg.

Stray, C. (1998) *Classics Transformed: Schools, Universities, and Society in England, 1830–1960.* Oxford: Clarendon Press.

Suetonius (trans. H. M. Bird, 1997) *Vitae Claudius (Life of Claudius).* Ware: Wordsworth Classics.

Suetonius (trans. H. M. Bird, 1997) *Vitae Augustus (Life of Augustus).* Ware: Wordsworth Classics.

Tamm, B. (1970) *Neros Gymnasium in Rome.* Stockholm: Almqvist & Wiksell.

Tatton-Brown, T. (1977) 'Excavations in 1976 by the Canterbury Archaeological Trust', *Archaeologia Cantiana*, 235–244.

Taylor, R. (1997) 'Torrent or trickle? The Aqua Alsietina, the Naumachia Augusti, and the Transtiberim', *American Journal of Archaeology*, 101(3), 465–492.

Taylor, R. (2012) 'Rome's lost aqueduct', *Archaeology*, 65(2), 34–40.

Terrenato, N. (2001) 'Ancestor cults: The perception of Rome in modern Italian culture', in R. Hingley (ed.) *Images of Rome: Perceptions of Ancient Rome in Europe and the United States in the Modern Age.* Portsmouth, RI: Journal of Roman Archaeology, 71–89.

Thompson, F. H. (1954) 'The Roman aqueduct at Lincoln', *Archaeological Journal*, 106–127.

Tilley, C. Y. (1994) *A Phenomenology of Landscape: Places, Paths, and Monuments.* Oxford: Berg.

Tilley, C. Y. (2010) *Interpreting Landscapes: Geologies, Topographies, Identities.* Walnut Creek, CA: Left Coast Press.

Todd, M. (1989) 'The early cities', in M. Todd and T. F. C. Blagg (eds) *Research on Roman Britain, 1960–89.* London: Society for Promotion of Roman Studies, 75–91.

Tomlin, R. S. O. (1983) 'Non Coritani Sed Corieltauvi', *Antiquaries Journal,* 63(2), 353–355.

Toynbee, J. M. C. (1962) *Art in Roman Britain.* London: Phaidon.

Vance, N. (1997) *The Victorians and Ancient Rome.* Oxford: Blackwell.

van Driel-Murray, C. (1999) 'And did those feet in ancient time ... Feet and shoes as a material projection of self', in P. Baker, C. Forcey, S. Jundi and R. Witcher (eds) *TRAC98: Proceedings of the Eighth Annual Theoretical Roman Archaeology Conference, Leicester, 1998.* Oxford: Oxbow Books, 131–141.

Varro (trans. R. G. Kent, 1951) *De lingua Latina (On the Latin Language).* London: Heinemann.

Virgil (trans. D. West, 2003) *The Aeneid.* London: Penguin.

Vitruvius (trans. F. Granger, 1931) *De architectura (On Architecture).* London: Henemann.

Wacher, J. (1963) 'Cirencester 1962: Third interim report', *Antiquaries Journal,* 43(1), 15–26.

Wacher, J. (1975) *The Towns of Roman Britain.* London: Batsford.

Wacher, J. (1978) 'The water supply of Londinium', in J. Bird, H. Chapman and J. Clark (eds) *Collectanea Londiniensia: Studies in London Archaeology and History Presented to Ralph Merrifield.* London: LAMAS, 104–108.

Wacher, J. (1995) *The Towns of Roman Britain,* 2nd edn. London: Batsford.

Wait, G. A. (1985) *Ritual and Religion in Iron Age Britain.* Oxford: BAR Publishing.

Waldron, H. A. (1973) 'Lead poisoning in the ancient world', *Medical History,* 17(4), 391–399.

Wardle, A. (2008) 'Bene lava: Bathing in Roman London', in J. Clark J. Cotton, J. Hall, R. Sherris and H. Swain (eds) *Londinium and Beyond: Essays on Roman London and its Hinterland for Harvey Sheldon.* York: Council for British Archaeology, 201–211.

Warren, S. H. (1913) *The Opening of the Romano-British Barrow on Mersea Island, Essex.* Colchester: Essex Archaeological Society.

Weber, M. (1958) *The City.* Glencoe, IL: Free Press.

Weber, M. (1978) *Economy and Society.* Berkeley, CA: University of California Press.

Webster, G. (1983) 'The function of Chedworth Roman "Villa"', *Transactions of Bristol and Gloucester Archaeological Society,* 101, 5–20.

Webster, G. (1988) 'Wroxeter (Viroconium)', in G. Webster (ed.) *Fortress into City: The Consolidation of Roman Britain First Century AD.* London: Batsford, 120–145.

Webster, J. (1995) 'Sanctuaries and sacred places', in M. J. Green (ed.) *The Celtic World.* London: Routledge, 445–465.

Webster, J. (2001) 'Creolizing the Roman provinces', *American Journal of Archaeology,* 105(2), 209–225.

Weekes, J. (2011) 'A review of Canterbury's Romano-British cemeteries', *Archaeologia Cantiana*, 131, 23–42.

Weekes, J. and P. Seary (2011) *Canterbury Westgate Gardens Desk-based Assessment*. Canterbury: Canterbury Archaeological Trust.

Wessex Archaeology (2009) *Friars Wash, Redbourn, Hertfordshire, Archaeological Evaluation and Assessment of Results*. Salisbury: Wessex Archaeology.

West, S. (2012) Personal communication.

Wheeler, M. (1943) *Maiden Castle, Dorset*. London: Society of Antiquaries of London.

Wheeler, R. E. M. and T. V. Wheeler (1936) *Verulamium: A Belgic and Two Roman Cities*. London: Society of Antiquaries of London.

White, R. H. (1999) 'The evolution of the baths complex at Wroxeter, Shropshire', in J. DeLaine and D. Johnston (eds) *Roman Baths and Bathing: Proceedings of the First International Conference on Roman Baths Held at Bath, England, 30 March–4 April 1992*. Portsmouth, RI: Journal of Roman Archaeology Supplementary Series 37, 278–291.

White, R. H. and P. Barker (1998) *Wroxeter: The Life and Death of a Roman City*. Stroud: Tempus.

Whitwell, J. (1976) *The Church Street Sewer and an Adjacent Building*. London: York Archaeological Trust/Council for British Archaeology.

Wigley, A. (2002) *Building Landscapes, Constructing Communities: Landscapes of the First Millennium BC in the Central West Marches*. Sheffield: University of Sheffield.

Wildfang, R. L. (2006) *Rome's Vestal Virgins: A Study of Rome's Vestal Priestesses in the Late Republic and Early Empire*. London: Routledge.

Wilkes, J. J. (1999) 'Approaching Roman baths', in J. DeLaine and D. Johnston (eds) *Roman Baths and Bathing: Proceedings of the First International Conference on Roman Baths Held at Bath, England, 30 March–4 April 1992*. Portsmouth, RI: Journal of Roman Archaeology Supplementary Series 37, 17–24.

Williams, A. and S. S. Frere (1949) *Roman Canterbury: An Account of the Excavations in Butchery Lane, Christmas 1945 and Easter 1946*. London: Medici Society.

Williams, S. (2006) *93 Nettleham Road, Lincoln: Archaeological Investigation Report*. Lincoln: Pre-Construct Archaeology.

Williams, T. (2003) 'Water and the Roman city: Life in Roman London', in P. Wilson (ed.) *The Archaeology of Roman Towns: Studies in Honour of John S. Wacher*. Oxford: Oxbow Books, 242–251.

Willis, S. (1999) 'Without and within: Aspects of culture and community in the Iron Age of north-eastern England', in B. Bevan (ed.) *Northern Exposure: Interpretative Devolution and the Iron Ages in Britain*. Leicester: Leicester University Press, 81–110.

Willis, S. (2005) 'Samian pottery, a resource for the study of Roman Britain and beyond: The results of the English Heritage funded Samian Project. An e-monograph', *Internet Archaeology*, 17.

Willis, S. (2006) 'The later Bronze Age and Iron Age (first millennium BC)', in N. J. Cooper (ed.) *The Archaeology of the East Midlands. An Archaeological Resource Assessment and Research Agenda*. Leicester: Leicester University Press, 89–136.

Willis, S. (2007a) 'Roman towns, Roman landscapes: The cultural terrain of town and country in the Roman period', in A. J. Fleming and R. Hingley (eds) *Prehistoric and Roman Landscapes: Landscape History after Hoskins*. Macclesfield: Windgather Press, 143–165.

Willis, S. (2007b) 'Sea, coast, estuary, land, and culture in Iron Age Britain', in C. Haselgrove and T. Moore (eds) *The Later Iron Age in Britain and Beyond*. Oxford: Oxbow Books, 107–130.

Willis, S. (2013) Personal communication.

Wilmott, T. (1982a) 'Water supply in the Roman city of London', *London Archaeologist*, 4(9), 234–242.

Wilmott, T. (1982b) 'Excavations at Queen Street, City of London, 1953 and 1960, and Roman timber-lined wells in London', *Transactions of London & Middlesex Archaeological Society*, 33, 1–78.

Wilmott, T. (1991) *Excavations in the Middle Walbrook Valley, City of London, 1927–1960*. London: London & Middlesex Archaeological Society.

Wissowa, G. (1912) *Religion und Kultus der Römer*. Munchen: Beck.

Witcher, R. (1998) 'Roman roads: Phenomenological perspectives on roads in the landscape', in C. Forcey, J. Hawthorne and R. Witcher (eds) *TRAC97: Proceedings of the Seventh Annual Theoretical Roman Archaeology Conference, Nottingham 1997*. Oxford: Oxbow Books, pp. 60-70.

Woodward, P. J. (1993) 'The discussion', in P. J. Woodward, S. M. Davies and A. H. Graham (eds) *Excavations at the Old Methodist Chapel and Greyhound Yard, Dorchester, 1981–1984*. Dorchester: Dorset Natural History and Archaeological Society, 351–380.

Woodward, P. J. and A. Woodward (2004) 'Dedicating the town: Urban foundation deposits in Roman Britain', *World Archaeology*, 36(1), 68–86.

Woodward, P. J., S. M. Davies and A. H. Graham (eds) (1993) *Excavations at the Old Methodist Chapel and Greyhound Yard, Dorchester, 1981–1984*. Dorchester: Dorset Natural History and Archaeological Society.

Wright, R. P. (1948) 'Roman Britain in 1947: I. Sites explored; II. Inscriptions', *Journal of Roman Studies*, 38, 81–104.

Wright, R. P. (1958) 'Roman Britain in 1957: I. Sites explored; II. Inscriptions', *Journal of Roman Studies*, 48(1/2), 130–155.

Wright, T. (1872) *Uriconium; A Historical Account of the Ancient Roman City, and of the Excavations Made Upon Its Site at Wroxeter, in Shropshire, etc.* London: Longmans, Green & Co.

WWF (2014) 'The state of England's chalk streams', WWF UK, available at www.wwf.org.uk/updates/state-englands-chalk-streams.

Yates, D. and R. Bradley (2010) 'Still water, hidden depths: The deposition of Bronze Age metalwork in the English fenland', *Antiquity*, 84(324), 405–415.

Yegül, F. K. (2010) *Bathing in the Roman World*. New York: Cambridge University Press.

Yule, B. (2005) *A Prestigious Roman Building Complex on the Southwark Waterfront: Excavations at Winchester Palace, London, 1983–90*. London: Museum of London Archaeology Service.

Zajac, N. (1999) 'The thermae: A policy of public health or personal legitimation?', in J. DeLaine and D. Johnston (eds) *Roman Baths and Bathing: Proceedings of the First International Conference on Roman Baths Held at Bath, England, 30*

March–4 April 1992. Portsmouth, RI: Journal of Roman Archaeology Supplementary Series 37, 99–107.

Zant, J. M. (1993) *The Brooks, Winchester 1987–88: The Roman Structural Remains*. Winchester: Winchester Museum Service.

Zienkiewicz, J. D. and D. Allen (1986) *The Legionary Fortress Baths at Caerleon*. Cardiff: National Museum of Wales.

Index

Page numbers in *italics* denote figures.